*The Chicago Guide to Landing a Job in Academic Biology*

# THE CHICAGO GUIDE TO LANDING A JOB IN ACADEMIC BIOLOGY

*C. Ray Chandler, Lorne M. Wolfe, & Daniel E. L. Promislow*

The University of Chicago Press   CHICAGO AND LONDON

C. RAY CHANDLER is a professor of biology at Georgia Southern University.
LORNE M. WOLFE is a professor of biology at Georgia Southern University.
DANIEL E. L. PROMISLOW is a professor of genetics at the University of Georgia.

The University of Chicago Press, Chicago 60637
The University of Chicago Press, Ltd., London
© 2007 by The University of Chicago
All rights reserved. Published 2007
Printed in the United States of America

16  15  14  13  12  11  10  09  08  07        1  2  3  4  5

ISBN-13: 978-0-226-10129-3 (cloth)
ISBN-13: 978-0-226-10130-9 (paper)

ISBN-10: 0-226-10129-0 (cloth)
ISBN-10: 0-226-10130-4 (paper)

Library of Congress Cataloging-in-Publication Data

Chandler, C. Ray.
The Chicago guide to landing a job in academic biology / C. Ray Chandler,
Lorne M. Wolfe, and Daniel E. L. Promislow.
p.  cm.
Includes bibliographical references.
ISBN-13: 978-0-226-10129-3 (cloth : alk. paper)
ISBN-10 0-226-10129-0 (cloth : alk. paper)
ISBN-13: 978-0-226-10130-9 (pbk. : alk. paper)
ISBN-10: 0-226-10130-4 (pbk. : alk. paper)
1. Biologists—Employment.
I. Wolfe, Lorne M. II. Promislow, Daniel E. L. III. Title.
QH314.C43   2007
570.71′1—dc22
2006029907

# CONTENTS

# PREFACE

After years of watching otherwise excellent job candidates fail to get job offers because of avoidable errors and wanting to make sure that his own students avoided these mistakes, Ray Chandler decided to write a paper for a widely read biology journal about the dos and don'ts of the job search. He enlisted Lorne Wolfe, and together they wrote a first draft. Lorne felt pretty good about the manuscript and gave it to Daniel Promislow for comments when he was visiting Lorne in Statesboro. After reading the paper, Daniel said to Lorne, "Nice idea, but it shouldn't be a paper. It should be a book."

And so was born *The Chicago Guide to Landing a Job in Academic Biology*.

There are many books out there on the academic path. You may have already seen, for example, *The Chicago Guide to Your Academic Career* (Goldsmith, Komlos, and Gold, 2001). This is a fine book that provides you with a reference not only for the job search, but also with much advice on how to write a thesis and what to do once you become an assistant professor.

We have decided to take just one aspect of this process, the job search, and distill it down to the essentials. We have deliberately written a book that is short, easy to read, and that we hope might even bring some levity to an inherently stressful process.

We have written this book for a broad audience, from undergraduate students thinking about going to graduate school, to those of you who have been offered one or more jobs and are about to start negotiating start-up packages and salary. If you are a professor who already has an academic job in biology, we hope that this book might serve as a useful resource for ways to help your own students and postdocs.

For those of you just starting out, you may be feeling just a little bit nervous, with lots of unanswered questions. Is academia right for me? Should I teach a course during grad school just for the experience? Should I accept that job offer from a school whose start-up package consists of a manual typewriter, two pairs of scissors, and a $10 gift certificate for the local hardware store?

There are so many decisions to make as you wend your way along this path. We hope that this book helps you figure out the answers that are right for you. *The Chicago Guide to Landing a Job in Academic Biology* does not provide a magic formula to guarantee that you land the job of your dreams, but it does offer a lot of common sense. In some cases, you may find yourself thinking, "Oh, so *that's* how it works." In others, the book may simply confirm what you already thought. And occasionally, you may disagree with our advice. That's okay too. The most important thing is that the path you pursue feels like the right one for you. If you go forward confident in your knowledge of how all this works, you will be more relaxed. And once you relax, you can start to have fun with this entire process.

Whether you are now at a conference and reading through this book in your hotel room, or sitting in your office taking a break from writing a manuscript, or on vacation (Hey! Don't think about work *all* the time!), we hope this brief book helps you figure out where you would like to end up as you pursue an academic career, and how to get there. We look forward to hearing your own stories.

# ACKNOWLEDGMENTS

*From so simple a beginning . . .*   CHARLES DARWIN

This book would never have come to fruition if it were not for the enormous help that we received throughout the entire process.

The quality of this final product was improved immensely by comments from Sheri Church, Lynda Delph, Steve Hudman, Leslie Rissler, Deborah Roach, and several other anonymous individuals. We were pleased to see that our colleagues put as much rigor into reviewing a book manuscript as they do our research publications! Kris Boudreau provided invaluable advice on the general outline of the book and the encouragement of a true veteran. We thank Matt Hahn for his willingness to share his research statement with us. Carol Froehlinger contributed comments and some of her own insightful words to the last chapter.

As scientists, we are used to having to generate new data every time we carry out an experiment. When we sat down to write this book, we discovered that there are vast resources available for students and postdocs thinking of pursuing an academic career. First and foremost, we discovered an abundance of high-quality online sources, which we will refer you to where appropriate. We are also grateful to John Wares, who shared the UC Davis PopBio "Red Binder" with us, and to the many authors who have written books like this in the past. If we are successful in reaching a new audience, it is in no small part thanks to the foundation created by past authors. We also benefited enormously from many anonymous and signed responses to our survey on the academic job-search experience. The quotes that we have placed in the text have been kept anonymous, as we wanted people to feel free to speak openly about their experiences.

Coauthoring a book like this is a bit like playing the surrealists' game of Exquisite Corpse, with each of us having his own style and set of ideas. Fortunately, our editor, Christie Henry, was able to handle three different personalities. Christie was instrumental in guiding this project to fruition, with encouragement, good humor, and thoughtfulness. We would never have met Christie were it not for the generosity of Michael Fisher of Harvard University Press, and we are grateful to him. Erin DeWitt provided impeccable copyediting.

We are indebted to Stefanie Ogawa for her uncanny ability to illustrate our words with her perfect vision, style, elegance, and fabulous humor.

We dedicate this book to all those faculty who served as our mentors and advisors during our academic training and development. During the course of our own job searches, we have each benefited from many friends, colleagues, and mentors, both formal and informal. With apologies to those many people whom we leave off this list, we do want to specifically acknowledge Spencer Barrett, May Berenbaum, Jim Curtsinger, Mark Gromko, Paul Harvey, Lynne Houck, Ellen Ketterson, Susan Mazer, Bob Montgomerie, Val Nolan, Bob Rose, John Rotenberry, Ruth Shaw, and Steve Vessey.

Finally, Ray thanks Michelle Cawthorn for her support and patience during the rigors of a long dual-career job search. Daniel thanks Julia Leong for her creative inspiration and for her understanding during the busy times. Lorne thanks Janet Burns, who has been a great partner along the road from postdoc to tenure, and beyond.

# 1

# The Academic Job Market

Doug and Holly work in the Department of Biology at a medium-sized public university in the Midwest.

Holly thrives in her department. In large lecture classes, her students are wowed by her energy and creativity. In her seminar classes, students feel like they are not just learning what scientists think, but how to think like scientists. Holly loves the process of scientific discovery and has made a point of including undergraduates in just about every research project in her lab. She loves working with the students and has been so productive that she was recently named editor of a major journal in her field. She also runs the master's program and has managed to recruit some excellent graduate students. Despite her busy life as teacher, researcher, administrator, and editor, Holly also manages to find time to relax with her family and play the accordion in a polka band! She has the perfect job (even if her taste in music leaves something to be desired).

Doug wasn't sure about taking an academic job at a school without a

Ph.D. program but liked the idea of living close to where he grew up. And without any ideas for alternative careers, he decided to give it a try. Doug joined Holly's department three years ago with great promise, having spent four years as a postdoc in a very productive biochemistry lab. Doug was amazed by just how different having an academic job was from life as

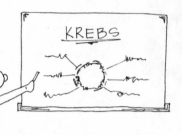

a postdoc, where all he did was research. Doug resents the fact that he has to teach, because it interferes with his research. His students sense this, and some of his courses have failed to attract enough students to justify their existence. Doug resents his department head because of the burden of the service tasks he has been assigned, which he thinks only interfere with his research. Of course, despite working late nights and weekends, he doesn't seem to get much research done, because of the burden of his teaching and service duties. Doug is worried about tenure, and rightly so.

Bruce flew in from Boston to interview for a job in Doug and Holly's department. During breakfast with the graduate students, he surprised them by telling them that he couldn't imagine living in the Midwest. In Bruce's seminar, he managed to combine a disinterested tone with an unfocused talk that gave no indication of his future direction.

Holly's happy, Doug's disgruntled, and Bruce is unemployed.

\*

This story nicely summarizes one of the major hurdles in the academic life of a biologist: getting an academic job in which you can be a happy and productive professional. What do Holly and Doug have that Bruce doesn't? Why is Doug unhappy in exactly the same job environment in which Holly thrives? The answer is complex because there is no single path in academia. For some, the road is straight and narrow. You might have already decided by elementary school that someday you would be working as an academic biologist—yes, some of us are that geeky. For others, the path is rather more unpredictable, with time spent in the Peace Corps, perhaps, or maybe a few years laying bricks (nothing like real work to drive you into academia). Some find academia enormously satisfying, while others never seem to be quite happy with the academic life. The three of us writing this book have followed our own, quite different paths to a faculty position in academia.

Ultimately all the pieces of the above story, and our own real lives as biologists, intersect at the one experience that all academics share—the job search. The outcome of the search determines whether or not you get a job (Holly vs. Bruce) and whether you wind up in a university that makes you happy or miserable (Holly vs. Doug). Surprisingly, during our graduate training, we give little thought to this phase in our academic careers on which so much depends and that nobody can avoid. Remarkably, biologists who spend ten years laboring over a dissertation or weeks assessing the merits of a single statistical test will exhaust their "training" in job-search skills in a single ten-minute chat with their advisor. This casual approach is inadequate because the process of searching for and obtaining an academic job at a modern university is challenging and complex. Furthermore, the amount of time and money invested in academic training is staggering; much is at stake.

If you have picked up this book, you have probably been in school for at least fifteen years, and maybe even distressingly close to twenty-five or thirty years. Pursuit of an academic career in the biological sciences requires more training than your average surgeon or NASA astronaut. So it is ironic that one of the most critical stages in this career—namely, following graduate or postdoctoral work with a successful job search and interview—is one for which graduate students and postdocs receive little or no formal training. All too often, after years of detailed course work, research training, and firsthand research experience, students are launched into the job market with only cursory guidance on how to search and interview successfully for a job consistent with their goals and abilities. The most common wisdom seems to be "publish a lot and ask about start-up." Sound advice, but hardly enough to cover the vagaries of searching for a position in a complex academic job market. And it is a complex market—far more complex than the simple "big school" versus "small school" dichotomy that most applicants use to characterize their job prospects.

It is our belief, based on our own—eventually successful—job searches and on our experience on search committees in recent years that most candidates for academic positions would benefit from specific guidance concerning (1) appropriate professional development during graduate training, (2) the search for an appropriate job opening, (3) the mechanics of applying for an academic job, and (4) the strategies for a successful interview. Although any successful search is predicated on a strong résumé (publications, good teaching experience), there are so many good competitors in today's job market that skills associated with the search, application, and interview process can make the difference between a job offer and a rejection letter. Thus, the purpose of this book is to provide formal guidance in developing these skills. This book will help you navigate the tricky terrain of an academic job in the sciences, from the first steps upon entering a graduate program, to the final start-up negotiations when you get that long-awaited job offer, and each step in between. We cannot guarantee you a job by the time you finish the last chapter of this book, but at least you will avoid the oft-repeated mistakes that we see in our role as search committee members.

## Ivory Tower or Well-Guarded Fortress?

Like an academic campfire story designed to scare graduate students, every graduate program has lore about the mysterious postdoc in the lab down the hall who has been searching for a job since the Reagan administration. Thus, landing an academic position sometimes seems about as probable as winning the Powerball Lottery. And just like the real Powerball, there are a lot of other hungry graduate students and postdocs waiting in line to buy tickets. How many? Currently in the United States, there are almost half a million graduate students, 30,000 postdoctoral researchers, and 170,000 faculty in science and engineering (National Science Foundation). About 50,000 new graduate students enroll in Ph.D. programs every year. All of these graduate students looking for academic jobs may sound daunting. But in fact, this is a perfect time to be preparing for an academic job in the

sciences. Over the course of just one month at the height of the autumn job-hunting season, several hundred faculty positions in the biological sciences are advertised in the pages of *Science* and *Nature*. For those interested in positions with a greater emphasis on teaching, we found many hundreds of jobs in biology in a single fall issue of the *Chronicle of Higher Education*. Of course, within narrower fields, there will be fewer jobs, but there will also be fewer competitors.

Table 1.1. Carnegie Foundation for the Advancement of Teaching classification of colleges and universities (2000)

| Carnegie Classification | Number of Schools |
| --- | --- |
| Doctoral/research universities—extensive | 150 |
| Doctoral/research universities—intensive | 112 |
| Master's (comprehensive) colleges and universities I | 490 |
| Master's (comprehensive) colleges and universities II | 127 |
| Baccalaureate colleges—liberal arts | 212 |
| Baccalaureate colleges—general | 308 |
| Baccalaureate/associate's colleges | 50 |
| Associate's colleges | 1,639 |
| Specialized institutions | 740 |
| Tribal colleges | 27 |

It turns out that for demographic reasons alone, the first decades of the twenty-first century are a great time to be looking for a job. As baby boomers reach their fifties and sixties, retirement rates among academic scientists have been increasing (Smaglik 2001). And as the children of those baby boomers hit college age, college enrollment is on the rise, leading to the need for more faculty. According to statistics from the Bureau of Labor Statistics at the U.S. Department of Labor, we can expect continued growth in the number of new faculty positions for several years. Of course, although we often think of academia as an ivory tower, we are not sheltered from the winds of economic change, and change they will. We do not claim to be long-term economic forecasters, but the academic job market does look pretty good if you are thinking about applying for an academic job some time in the next few years.

So why are so many people spending decades training for a job in academia? Are we really having that much fun in here? It would not seem so at face value. There are easier jobs. There are jobs with shorter workweeks. There are jobs that pay more. Nevertheless, we have chosen the academic side of science, and we are shameless boosters.

Before we discuss the costs and benefits of an academic job, let's look at the academic landscape more closely. A newly minted Ph.D. or postdoc will look out onto an academic landscape composed of a remarkable variety of colleges and universities. The Carnegie Foundation for the Advancement of Teaching classifies these schools based on the types of degrees awarded and the range of programs (table 1.1). Of most interest to those seeking an academic job in biology—and following the breakdown used by *U.S. News* in their annual rankings of colleges and universities—there are doctoral universities (approximately 260 in the United States), master's colleges and universities (approximately 617), liberal arts colleges (approximately 212), and comprehensive bachelor's colleges (approximately 358). Thus, universities granting appreciable numbers of doctoral degrees like the institution where you trained as a Ph.D. student represent less than 20 percent of the schools to which you might apply for a job. (We are ignoring in our calculations associate's colleges, specialized institutions, and tribal colleges.) The take-home message is that much of the job market lies outside the type of university where most graduate and postdoctoral training happens.

The Carnegie classification from doctoral to bachelor's roughly defines a continuum of jobs, ranging from those that emphasize research, publications, graduate training, and external grant funding to those that emphasize undergraduate teaching, perhaps with a focus on research as a teaching tool for the undergraduate classroom. If you take a job at a major doctoral institution, you can expect a light teaching load (often one course per semester or year, depending on your level of grant support). However, you will be expected to secure major external funding on a regular basis that is sufficient to fund your research and support students. You must publish several papers a year in good journals. The pressure to publish and obtain grant funding can be intense. At a bachelor's or liberal arts institution, there will be much greater emphasis on undergraduate teaching. At the extreme you might teach twelve contact hours per semester and be responsible for three different courses. You will be expected to interact regularly with undergraduates. Grants and research will usually be evaluated in light of teaching or undergraduate involvement, and there will be less pressure to obtain external grants or to publish. However, the pressure of juggling diverse teaching demands and many students can be intense. In the vast middle, you can find institutions with job descriptions that fall almost anywhere between these extremes.

Regardless of the institution, most academic biologists will be seeking a tenure-track assistant professor position. These are probationary positions that offer the chance for tenure after five to seven years. As an assistant professor, you will be responsible for the full range of academic responsibility: teaching, research, and service. However, there are other routes to permanent jobs inside the ivory tower. For those whose interest lies almost exclusively with teaching, many universities hire instructors and laboratory coordinators whose sole responsibility is to teach (often large introductory courses) or to coordinate introductory labs. These can be excellent jobs,

---

**Box 1.1. General costs and benefits of the academic life**

**Benefits**

Stimulating working environment with opportunity for lifelong learning

Good opportunities for travel

Casual working environment; every day is casual Friday

Rewards creativity; potential for international recognition

Good opportunity for advancement; tenure

Diverse, intellectually challenging work

Rewarding opportunities for teaching and mentoring

Work hours are flexible; considerable freedom to direct your own activity

**Costs**

Relatively low pay

Pressure for grants and publications is stressful

Students can be demanding

Can require long hours

Lack of public understanding

Relatively fixed timetable for advancement

---

with considerable opportunity to direct introductory programs, design curricula, and create laboratory experiences. At the other extreme are research associate positions in which the only responsibility is research. Although these jobs may technically be permanent, they are typically funded by soft (i.e., grant) money. In some cases, this money comes through the university, department, or research institute. In other cases, a research associate simply has a mailing address and a hunting license to obtain his or her own grant funding (including salary).

The most common route into academia is a position as a college professor, and this is the job search that we largely have in mind in writing this book. Of the many career paths one can follow, a faculty position stands out for the intellectual challenges and diverse rewards it can provide. Of course, the job comes with significant sources of stress and frustration as well ("The dean put me on the Faculty Roles and Rewards Committee?"). We list our assessment of the costs and benefits of the academic life in box 1.1. For us and apparently for thousands of others as well, the benefits far outweigh the costs. Part of what makes this job so exciting is the many different aspects of life as a scientist. As an academic scientist, you will become not only researcher and teacher, but also writer, editor, counselor, advisor, administrator, manager, and more. But unlike in industry, where one's chances to affect change may be fairly limited, academia offers faculty members the chance to play a central role in the life of the institution and those who are a part of it, from students to administrators. It is a job worth searching for.

### Finding a Job in Ten Easy Steps

Chances are if you are reading this book, you have already decided to seek an academic job or are already on the job market. A number of important considerations go into this decision, but we will talk more about this later. For now, suffice it to say that we can help with your search. Our perspective on the academic job search is shaped by our own graduate training at various-sized institutions and by over thirty years of cumulative postgraduate academic experience. We have also been influenced strongly by our participation in many tenure-track job searches over the past decade. During these searches, we have been surprised by the tremendous variation in the quality of the presentation of applications, the remarkably poor performance of some candidates during phases of the interview that they should have been able to anticipate and prepare for, and the rather superficial level at which many candidates have considered their academic career options and goals. Our inescapable conclusion is that many good job candidates are handicapping their success on the job market because of correctable mistakes in career planning, preparation of applications, and interview technique.

Other books have provided advice for carrying out a successful job search in academia (e.g., Goldsmith et al. 2001); in some ways we echo these excellent recommendations. However, our discussion takes a more integrated view of the job-search process. We explicitly link events from selection of a graduate program to interviewing as components of a single unified process. This process is characterized by the need to consider the range of career options in academia and which of those options is most consistent with your skills and aspirations. We also offer uniquely practical advice for the academic job search over the course of the next ten chapters; we will explain that there are simple steps you can follow to avoid blunders and permit your strengths to shine. Here is a quick overview of where we are headed.

## Get on the Right Road

Chapters 2–4 describe some important foundational work that you should do—from choosing your graduate program to life as a senior postdoctoral researcher—that will enhance your ability to find and obtain the job that is right for you. "The job that is right for you" is not just a tired cliché. The title "college professor" hides tremendous variation in job demands, skills required, rewards, and stresses. It is up to you as a graduate student or a postdoc to become familiar with this variation and to figure out where your strengths and career goals fit in. Remember, when choosing *among* jobs, you have much flexibility to choose one that allows you to emphasize your strengths (be they teaching, research, or service). Academia offers much less flexibility *within* jobs to identify and emphasize your strengths (flagship research institutions are not going to be very receptive to your decision that research is not your thing after all and that you want to teach more).

Chapter 3 will provide a road map for your early job preparation. We describe the kinds of things you should do early in your career to ensure that you have everything you need when you open *Science* to begin to search for that first job. We discuss important issues related to research, grants, teaching experience, role models, and overall career training. Chapter 4 will help you target a job search toward the sort of university that will maximize your potential, your job satisfaction, and your chance of reaching your ultimate career goal. Do you want to be the high-profile researcher who is hesitant to put any more time into teaching or service than is absolutely necessary? Maybe you aspire to be the renowned teacher whose courses are always packed and who loves to inspire students through well-crafted lectures. Or—bless your heart—maybe you want to be a graduate coordinator or department head who single-handedly turns a mediocre graduate program into a top-flight training program with international recognition.

An important premise of these three chapters relates to the structure of most graduate programs in biology. These programs excel at research training. There is no facet of the research enterprise for which a graduate student in biology cannot readily find assistance and advice. Unfortunately, these same graduate programs are woefully inadequate in providing career training, in the sense of exposure, training, and unbiased advice about the full range of job options within academia. Chapters 2–4 will help you take control of this facet of your job preparation.

## The Package

Once you have settled on the sort of job you are going to apply for, you need to put together a first-rate application. This may seem to be an easy task. How hard can it be to stuff a CV into an envelope? In reality, an academic job application is a complex document that must convey large amounts of  information to a difficult audience (faculty members looking for reasons to reject applicants from a candidate pool). Furthermore, we can attest to the fact that many job applications are composed of weak, poorly prepared documents. Chapter 5 will help you avoid becoming an example of a bad application for a future edition of this book.

Your job application will include a cover letter, statements of research and teaching interests, a carefully chosen list of references, and most importantly, a curriculum vitae with your educational experiences and publications. To be a successful candidate, you *must* craft these documents carefully. Happily, you can observe simple strategies to show yourself in the best possible light to the search committee. Remember, it is through this carefully assembled package that a search committee will first get to know you. From the small details of your career to the big picture of how you will excel in the advertised position, in chapter 5 we will show you how to use the application package to put your best foot forward.

## Face Time

If everything goes well, a month or two after you submit your application, you'll be making travel plans. You have convinced a department that you are worth getting to know firsthand, and they put you on the short list of candidates to be brought in for an interview. You will need to be on your best behavior even before you get there. One prospective candidate we met, on being informed by our department head that he had been put on the short list, asked if perhaps there had not been some mistake. Chapter 6 will help you by covering the kinds of things you should be doing to prepare for an interview. Do not underestimate how important these weeks before the interview are to your success. Once you arrive on campus, you will be in for a couple of days of intense, but exciting, interactions. You will want to give a smooth, prepared, and confident performance, and chapters 7–9 will take you step-by-step through the whole process, including your job seminar, faculty interviews, roundtables, meetings with graduate students, meals, and even social hours.

### Closing the Deal

Should that coveted offer come through, you have not quite finished your job (nor have we finished ours). It still remains to close the deal, to decide on issues like salary, lab space, start-up money, and so on. (And the decision is, in fact, partly up to you.) For scientists, graduate school teaches us little or nothing about the business side of the job hunt (nor about the business of running a lab—for that, you might want to have a look at other books, such as *Making the Right Moves* from the Howard Hughes Medical Institute—http://www.hhmi.org/grants/office/graduate/lab_book.html—or *At the Helm: A Laboratory Navigator* by Kathy Barker). There is much to gain at this stage of the process. There is also much to lose. How well you negotiate at this point will set the stage for the next few critical years of your career. Once you have a job offer in hand, the rest is up to you. Up to you, that is, and the network of supporters that you will begin to build throughout the university before you even arrive. With the help of friends, colleagues, and the Internet, you will need to determine what reasonable levels are for your salary, start-up package, teaching load, administrative duties, lab space, and so forth. And you will need to learn how to negotiate for all of these. Chapter 10 will guide you through these negotiations.

### "But Things Aren't So Simple . . ."

Finally, chapter 11 offers some thoughts (concrete advice is difficult) on some of the wild cards you might find in your hand as you embark on an academic job search. This chapter will help you better understand how to deal with some of the challenges related to two-career couples, family issues, inappropriate questioning, and discrimination. These are sensitive but important issues. Three white males cannot hope to provide advice with "street cred" for all these issues, but we at least try to bring the issues into focus for the job seeker.

## Summary

We are confident that by reading this book, you can increase your success in searching for an academic job. But we also appreciate that there is no single way to ensure that you receive a job offer from your first-choice institution. What works best for you will depend on your own strengths and experiences. We do not profess to have a monopoly on the absolute truth when it comes to searching for an academic job. We offer some time-tested solutions, but not the only solutions.

We hope that you find this book a useful guide. Coupled with suggestions from the experts around you—your mentor, other postdocs on the job market, recent hires in your department—this small book should set you on your way toward a rewarding academic career.

**2**

## Choosing a Graduate Program

If you have picked up this book, you may already be on the job market or nearing that stage in your career. However, some of you are just starting out as undergraduates, perhaps feeling a bit like it's your first day at school. Fortunately for you, the ideal time to begin planning for an academic job in biology is before you even enter a graduate program. Because your graduate training will provide the core qualifications that will land you that first academic job, the choice of a graduate program is a first critical step in the job hunt. Fortunately, there are a lot of resources to help in the hunt for a graduate program. Universities produce elaborate web pages and colorful brochures touting the strengths of their graduate programs. Scores of publications review and compare programs, and *U.S. News* produces a much-talked-about ranking of colleges, including graduate programs, every year. You might just doze off before you assimilate all the quantitative and comparative information that is available. However, from a career perspective, much of the data in brochures and rankings is of limited value. The factors that determine whether a graduate program will put you on a successful path to an academic job go well beyond a university's reputation and are rarely discussed in the usual graduate brochure or web page. The purpose of this chapter is to provide some practical advice on choosing a graduate program from the perspective of what you will need when it comes time to apply for that academic job.

What does a successful job applicant need from their graduate program? They need a program that provides solid advanced training in biology, the opportunity to conduct productive independent research, the ability to gain a wide range of relevant career experience (teaching, grantsmanship, service), and the chance to develop a supportive network of professional colleagues. The trick is to pick a program that will meet these needs, and a *U.S. News* ranking is unlikely to be very informative about these issues. We suggest the following guidelines.

## The University

### So Many Choices

If you are in the early stages of preparing for a career in academia, select a major respected university for your graduate training. There is no escaping the fact that a big-name school will garner extra attention, and many search committee members will be looking for applicants with solid pedigrees. For many committee members, the Ph.D. institution provides the first bit of useful information about an applicant. Major institutions also deliver a host of perks: world-class libraries, modern facilities, large pools of collaborators, broad departmental expertise, and abundant grant funding. There is no question about it; a "big" university can provide all the tools for success.

However, so many factors go into successful graduate work and professional training that the size or reputation of the university alone should *not* be the sole determinant of where you do your graduate work. Assume that this guideline carries an "all other things being equal" clause. If there is a range of schools that are equal in the degree to which they fit your requirements for a graduate program, then the better-known program will be a good career move. If big-name programs do not meet your needs, then do not pick those programs on name recognition alone. Consider three important facts. First, the productivity of your graduate career will be largely influenced by interactions between you and your advisor, committee mem bers, and lab mates. Your working relationship with this circle of mentors and colleagues will determine your personal and professional satisfaction with graduate school, and this relationship explains far more variation in career success than university reputation. Obviously, larger doctoral institutions do not have a monopoly on smooth, productive relationships among students, faculty, and peers. Second, a tight job market and strong growth in many midsize universities mean that good faculty and strong programs are more widely distributed than ever before. Productive scientists, including leaders in particular fields, can now often be found at a smaller school with a limited range of graduate programs. Do not overlook these programs just for the sake of a prestigious name. Finally, your success in competing for jobs will ultimately be based on your publications, grants, teaching background, and other professional activity. No program in the world has a sufficient reputation to place their graduates in academic positions without some authentic track record on the part of the candidate. So keep the reputation of a graduate program in perspective.

## The People

### Picking a Prof

Carefully evaluate your potential advisor before selecting a graduate program. No individual will have a greater influence over the success of your graduate work than your major advisor. He or she is arguably the single most important person in your professional career. A great advisor can help you to develop the skills and experience as well as provide the personal contacts that will set you on the path to an excellent career in biology. Therefore, we recommend that you spend a significant amount of time evaluating potential advisors, including face-to-face meetings. After all, choosing an advisor is like choosing a doctor—what suits one person perfectly may not work for another.

> *"The best application I ever received from a prospective graduate student was on bright orange paper. It caught my attention (of course) and was very well put together. When I finally interviewed the applicant, I asked her what she was thinking by doing that, and she replied that she did not want to work with a professor who was so stodgy that they would think an orange application was a bad idea."*

In many fields of biology, students are more likely to enter a program without a specific advisor in mind. Students will first carry out several rotations in different labs during their first year, after which they will choose a specific major professor. In this case, as you are choosing graduate programs, determine if there are two or three labs that could potentially meet your needs.

In either case, the search for a graduate program really becomes the search for excellent potential major advisors. Beginning well before you hope to start a graduate program, you should use publications, word of mouth, web pages, and professional meetings to identify individuals who have research programs that fit well with your interests. Seek input from other faculty and from peers about the person you are interested in. Remember, you are going to spend several years with this person, often under highly stressful situations. There is much you need to know.

It is appropriate to contact potential advisors before you even apply to a program. If you are able to impress a prospective advisor, he or she can often help to ensure that you are admitted into the program. In some programs, your admission will depend on a faculty sponsor. In most cases, before you are formally admitted, you will be invited for an interview. (If you have been invited to four or five schools, this may become your first experience of the close relationship between academic careers and frequent-flier mileage programs!) The interview allows the school to which you have applied to assess your suitability. But even more than that, it is an excellent opportunity for you to have your own questions answered (box 2.1).

- First and foremost, is this a person with whom you are compatible? You must be able to interact smoothly enough to foster a productive professional relationship. You do not necessarily have to interact socially with your advisor or hang out together at the mall; you don't even really have to like your advisor. However, you must have mutual respect and the ability to communicate clearly and productively.
- Second, you will want to explore what your research options are with this advisor. Can you work on a project that interests you? Will you have an adequate level of independence in choosing or carrying out the project, or will you be given a prefab cookie-cutter project with aims and approaches that were developed without much of your own input? If you are independent

**Box 2.1. Good questions to ask a potential advisor when choosing a graduate program**

- Will you provide a project for me as part of a larger research program, or do your students design their own relatively independent thesis projects?
- How will my project be funded? Will I have the opportunity to develop my own grant proposals?
- How many students have graduated from your lab in recent years? What are they doing now?
- What kinds of internal support are available through the department and university?
- What are your goals for your lab over the next two to five years? Do you have any plans to move or retire in the next five years?
- Will I have the opportunity to teach or serve as a teaching assistant? Do graduate students ever have the opportunity to teach lecture sections?
- Do graduate students commonly pursue side projects or collaborate among themselves?
- What facilities relevant to my research are available in your lab? What is available for use within the department?
- What is your philosophy or approach to authorship on papers? If I am supported by your grant, will you automatically be an author on all manuscripts?
- Will you or the department fund or defray the costs of travel for meetings?

and strong-willed, make sure you select an advisor who will give you sufficient freedom to work independently. If you need daily guidance, be sure that your advisor will provide this.

- Third, will the advisor be fair and ethical in distributing credit for the research (including authorship)? It is a good idea to broach issues related to credit and authorship up front. If you will be working on part of the advisor's research program, be sure that you will have the opportunity to establish your own independent reputation.
- Fourth, will the potential advisor be at the institution for the duration of your graduate work? A move at a critical point in your graduate work can

be difficult (though many graduate students have survived such a move). We even know of a case where an advisor changed universities over the summer without telling a graduate student! Of course, the best academics are likely to be in demand by other institutions—a move may be a reflection of his or her outstanding qualities. If an advisor does move, find out if you would be able to follow him or her to the new school. And determine for yourself whether you would want to do this.

· Fifth, is your advisor a seasoned veteran with an established international reputation and a large, extremely productive lab, or perhaps someone just starting out? There are obviously many benefits to working with someone who has already trained many students, but there are also advantages to joining the lab of someone who is not long out of graduate school. He or she will have a better understanding of your own needs, will likely spend more time in the lab working closely with students, and if younger, might be like the Energizer Bunny, with the high optimism and drive of a new faculty member.

· Finally, double-check the answers to these questions by looking into the track record of the potential advisor. Are they graduating steady numbers of students? Are these students happy? Are these students employed? Current and past students are excellent sources of information when evaluating a potential advisor.

### The Supporting Cast

Your advisor is important, but he or she is only the lead actor in a rather large cast that will determine your success as a graduate student. A good graduate program must offer a strong supporting cast. Does the department have the expertise to provide a strong committee? Are there enough faculty members in your discipline to ensure good intellectual interactions? Are there enough graduate students in your potential advisor's lab or in the department as a whole to have a strong peer group? If you do interdisciplinary work, are the staff and programs in related areas strong enough to support your graduate work? When you visit a graduate program to meet with a potential advisor, be sure to meet with some of the other faculty and take the time to chat with graduate students about their advisory committees and their interactions with faculty at large.

As you meet people and evaluate a program, realize that biologists are not always one big, happy family. It is not uncommon to find departments in which certain faculty (and often all the members of their labs) will not interact. Sometimes subdisciplines are at war over space and resources. Groups of graduate students can sometimes fragment into high school–like cliques. None of these things will necessarily compromise your ability to do graduate work, but they represent a significant risk. Dysfunctional departments will tend to create practical problems by denying you access to space or resources involved in a departmental dispute, and your ability to seek intellectual interaction freely across the department can be limited. At the extreme we know of cases where faculty members actively interfered with a graduate student's work as part of their feud with another faculty member. It is best if you look for and avoid these sorts of departments.

### Funding

Graduate funding options can affect your future job competitiveness. The dream scenario for many graduate students is an advisor who has major grant funding and is willing to pay a fat assistantship to fund several years of graduate work. Freed from financial worry and pesky teaching assistantships, the lucky graduate student can now concentrate on research. This is indeed a good deal. However, a full-ride assistantship on your advisor's grant does have some consequences for career training that cannot be overlooked. Do not enter a graduate program without giving some thought to the consequences of various funding options.

A principal goal of your graduate work is to do productive, cutting-edge research. A research assistantship from your advisor's grant can certainly facilitate good research. More research means you will be more competitive for jobs down the road. There is a significant risk to this strategy. If your research represents part of the grant obligations of the major advisor, it can be more difficult to establish your own independent research identity. When we serve on search committees, we see this lack of independence in two ways. Fully supported students often have little or no grant activity of their own, and during their research seminar they often come off sounding like research technicians, not independent investigators. We are not suggesting that you turn down grant-funded assistantships, but we are pointing out that you should be cognizant of how it might impact your ability to establish an independent research presence. Bear in mind that in many fields of biology, graduate students are quite dependent on their major professor's research program. If you expect to be in this situation, it will be especially important to find a postdoc where you can develop at least one independent line of inquiry.

The other traditional source of funding for graduate students is a teaching assistantship. In return for a stipend (and usually a tuition waiver), you will be asked to teach a minimum number of undergraduate laboratories. Although this teaching represents time away from your research, it is probably your first introduction to an important component of the vast majority of academic jobs. We believe that students should seek out a teaching assistantship during at least some portion of their graduate training. Remember, less than 20 percent of the universities out there are doctoral institutions. Many of the others will consider teaching assistantship experience the *minimum* level of teaching experience they want to see in applicants for a faculty position. If you are evaluating graduate programs, you need to know whether teaching assistantships are offered and how they are likely to figure into your support during graduate study.

> "Oddly enough, being a TA in graduate school probably did more than anything else to teach me about the time-management juggling act that is academia."

Finally, some graduate students are funded by fellowships, scholarships, or their own grant activity. Funding of this nature typically allows a student to work full-time on research, and the funding is competitive (often highly competitive). This sort of funding provides the same advantage as a research assistantship; however, there is far less potential to be overshadowed in your advisor's research program. Obtaining your own fellowship or grant is a big plus in a job application. This is the kind of independent professional activity that every graduate student should strive for because search committees expect to see evidence that an applicant can fund their own research. Spend some time assessing what kind of scholarship or fellowship opportunities a graduate program provides. Ideally, you would like to be in a program that is rich in internal funding opportunities, including funds for travel and professional development. However, also devote time to external grant sources (box 2.2). The Internet is a fantastic resource for information about graduate fellowships, broad-based grants (such as the NSF predocs), and grants specific to particular research fields.

Ultimately, a prospective graduate student should assess which of these funding options are available in a graduate program and how they are likely

---

**Box 2.2. Some selected online sources for information about grants and fellowships**

http://www.asee.org/fellowship/
http://biology.berkeley.edu/grants.html
http://chronicle.com/free/grants/
http://es.epa.gov/ncer/rfa/
http://www.fordfound.org/
http://www.grantsnet.org/
http://www.hhmi.org
http://www.library.wisc.edu/guides/Biology/jobs_funding.htm
http://www.nationalacademies.org/grantprograms.html
http://www.nationalgeographic.com/research/grant/rg1.html
http://nextwave.sciencemag.org/
http://www.nsf.gov/

---

to work together to fund a multiple-year graduate program. Regardless of the source, a graduate program must be able to provide continuity of funding that will ensure steady progress toward a degree. As we have described, that funding then has consequences for how you will be perceived by search committees when you apply for an academic job.

## Resources

### Physical Facilities

Your graduate program should provide all the necessary physical facilities for good research. Assuming that you have chosen an advisor whose research is compatible with your interests, the program in question should have the facilities to support this work. Larger doctoral institutions will typically have excellent research space and support facilities. Nevertheless, it will not hurt to make this assessment for your own particular needs. That beautiful greenhouse on the roof is of no use to you if there is too much competition for space or it costs too much (if you are funding your own work). As you consider smaller doctoral programs, you can sometimes find important limiting factors. In all cases, it is a good idea for a prospective student to make a mental checklist of the sort of facilities that are needed to work in their preferred research area. Don't be afraid to raise this issue with a potential advisor.

## Training Opportunities

Select a graduate program that provides the opportunity to train broadly. A major theme of this book is the importance of obtaining diverse professional experience (and an exposure to a range of alternative academic lifestyles) during your graduate training. We will talk more about this in the next two chapters, but when you choose a graduate program, you should explore what kind of flexibility and opportunities that program will provide in terms of training for the academic career *you* want. Will your advisor be flexible enough to allow you to attend a teaching seminar or to opt out of a research assistantship in order to teach for a semester? Does the department provide opportunities for students nearing the end of their graduate work to teach a lecture session of an introductory course? If you want to get involved in professional service (e.g., the student chapter of a professional society or the campus graduate organization), will your advisor and committee encourage or discourage you? Some advisors and programs expect their graduate students to be eating, sleeping (as long as it is not more than four hours a night), or in the lab doing research. If this is the intensity level at which you want to pursue research in an academic career, then this is the program for you. Other programs will provide greater space for the pursuit of teaching or other career interests. You need to seek out this kind of graduate program if you have your eye on a small liberal arts college in the future.

## A Final Note

Graduate school can be a fantastic experience, a chance to spread your intellectual wings, to surround yourself with like-minded people, to discover the joy and excitement (and, yes, sometimes the frustration) of scientific discovery. What's more, it's also a chance to discover the true challenges of multitasking, to figure out how to flourish despite the multiple demands of research, teaching, grant writing, service, and your personal life, which will sometimes make you wish that you had thirty-six hours a day and a few extra arms!

Clearly, there's a lot at stake in choosing a graduate school. But how do you get into your chosen graduate school in the first place? If you are an undergraduate reading this book, the advice is simple. Good grades and a high GRE score will clearly help (and can sometimes be the key to graduate fellowships). But even more than grades and GRE scores, research experience can make all the difference. Find an interesting and productive lab, and get involved early. Even if you are a freshman or sophomore just beginning your studies in science, consider volunteering in a lab. You may just be washing glassware, but you'll see how scientists work, you will have a chance to impress people in the lab with your excellent work ethic, and before you know it, you might just be on the way to carrying out your own independent experiments. And here's the critical factor: a graduate school application that includes a strong letter from a well-respected scientist for whom you have worked can be the most compelling element of a graduate school application.

**3**

*Prepare Early for Your Job Search*

If you ask a typical postdoc or recent Ph.D. what they have done to prepare for a job search, most will talk about the time they spent thumbing through the job ads in *Science* or *Chronicle of Higher Education*, polishing their seminar, or working to get manuscripts finished. Some more thoughtful individuals may point to the fact that they pursued doctoral training in the first place. A very few may mention choosing a research area that they thought would be marketable. While these are all necessary parts of preparing for a successful job search in academia, they are not sufficient. In fact, they are not even the most important when it comes to landing the *right* job.

The faculty ranks in academia are perhaps unique in the number of grumpy, disgruntled employees they hold. Any academic can tell story after story about colleagues who are routinely dissatisfied with their job as a faculty member. You know the one. Every conversation degenerates into a rant about the department, the university, the students, and so on. The gist of the story is usually that this faculty member is profoundly dissatisfied with their employer and the parameters of their employment. Sure, it's true that there are a lot of unhappy burger-flippers and cashiers out there, but they don't spend more than a decade training for their unrewarding job. Why would a new faculty member at the end of this protracted—even tortuous—training process ever be surprised, much less unhappy, with the job options they face? We believe the answer is ultimately a lack of preparation. Many, if not most, Ph.D.'s do not adequately prepare for, or even familiarize themselves with, the job options that are available. The most successful job searches—in the sense of getting a job that closely matches your strengths and aspirations—are those in which the candidate prepares early by ensuring that their training is appropriate to the type of demands that an academic job will actually place on them. This chapter will discuss this sort of preparation, and chapter 4 discusses how to target your search toward the appropriate job. With luck, this advice will steer you toward the happy place described by one of our colleagues at a large public university:

> "I love nearly everything about my job. No wonder lots of people want it (my job, that is). I can't think of a better way to spend my life. I interact daily with interesting, intelligent people and get to be creative. What I do is ever changing and therefore rarely boring. Plus, I'm good at what I do and that brings happiness."

When you get right down to it, a candidate who prepares early is one who begins as a graduate student and as a postdoc to take on many different tasks that will eventually make up your life as a faculty member. This means carrying out quality research, certainly, but also becoming involved in teaching, grantsmanship, and even service. All these activities will not only help you develop an impressive CV—they will also give you a good sense of what type of job, with what balance of activities, best suits your interests and abilities.

### Isn't It All about Research?

A famous hotelier, Charles Forte, was once asked what the secret was to the success of his investments. "Location, location, location," he is purported to have replied. Similarly, if a doctoral student gets any advice at all about early career planning, it is usually "publish, publish, and publish some more." As far as it goes, this is good advice. Publications and grantsmanship are the gold standard for documenting one's qualifications for a career in academia. Regardless of the kind of academic job you seek, a demonstration of research skills is an important part of landing a job. Fortunately, academic training is overwhelmingly focused on research and research training. No one receiving a Ph.D. in the sciences, no matter how little career planning they have done, will fail to receive extensive, hands-on experience with research. Thus, preparing for the research demands of an academic career is more about the quality of training and the resulting publications than it is a question of being exposed to research in the first place. Nevertheless, from the perspective of early career planning, there are several important points to consider.

First, as we have already described in chapter 2, where you go for graduate or postdoctoral training is an important first step on the long road to an academic job. We would only reiterate here that the careful consideration that you put into choosing a graduate school and mentor is also important when choosing where and with whom you will pursue postdoctoral training.

A second important point for early career preparation is to plan out a coherent program of research. This is an important step that should be thought about up front, not imposed on your work only as you begin to prepare your first job application. Realize that prospective

employers are not looking for brilliant publications in isolation. They want to see that you are pursuing a coherent research program with the promise of productive lines of inquiry in the future. We have seen many candidates who jump all the hurdles of the interview process only to fail at the very end because faculty deliberations raised questions about their research program or the coherence of the questions they were asking. Develop and think about these issues early in your training.

Third, make plans to establish your research independence early in your career. The huge expense associated with modern biology means that many—perhaps most—graduate students and postdocs work on research projects being funded by grants to advisors or supervisors. It is inevitable that a young scientist's research program is closely related to, or even dictated by, that of a more experienced scientist. This brings many perks in terms of funding, intimate training with respected scientists, and the opportunity to pursue questions that would not be available to an inexperienced young scientist. However, by the time you interview for an academic job, your audience will be looking for signs of your own independent research program. If they perceive that you have been funded solely by your supervisor and addressed questions generated solely by his/her work, using your supervisor's methodology, then there will be red flags concerning your application. In the years leading up to the job search, look for ways—even small ways—of establishing an independent research program. Ways of doing this include side projects, collaborations outside your home lab, introduction of novel methods into the lab in which you work, and leading your supervisor or peers into new research directions (i.e., you are the legitimate "author" of an independent line of research inquiry). Right from the start of your postdoctoral work, determine with your advisor what work you will be able to take with you when you start your first tenure-track job and what will need to stay behind. If you are paid by your advisor, as opposed to bringing in your own funding, your advisor will likely want you to work on problems that further the aims of the advisor's grant. But he or she will also recognize that you need to establish independence as you move forward.

These days, it is especially valuable to establish an independent record of grantsmanship during your graduate and postdoctoral years. We're not necessarily talking about landing your own NSF or NIH grants. There are scores of grant opportunities for young biologists that can help demonstrate to a search committee that you have the basic skills to obtain your own grants. These grants are often small, and they may seem like a waste of time to your advisor or supervisor. However, you need to be looking out for your own best interests in establishing an independent grant history. The list of possible grant sources is too long and changes too rapidly to provide an exhaustive list here (but see box 2.2 for some sources). To start down this road, explore the grants offered by private foundations, professional societies, museums, state and federal agencies, corporations, nongovernmental organizations, and so on. Independent grantsmanship is an area that will get careful attention from search committees, even at smaller schools.

### It's Not All about Research

Graduate programs in biology typically excel at providing research training. It often seems that every waking moment for a bleary-eyed doctoral student is committed to feeding the beast. While less-encumbered friends are heading out to a concert, the dedicated graduate student is wading through that 1884 monograph on molt patterns in warblers or running yet another gel. However, from the perspective of career training, we would also hope that our graduate programs provide a broad introduction to the range of academic jobs available to a new Ph.D. This would include open-minded discussion of the strengths and weaknesses of various job settings. Graduate students and postdocs would also profit from exposure to a range of academic role models. What better way to plan for an academic career than to see different kinds of practicing faculty members? Teaching would ideally be integrated into research training and be seen as a valuable part of career preparation. These are all critical steps toward ensuring that new Ph.D.'s are exposed to the full range of employment opportunities and permitted to train for the jobs that best suit their career goals.

Unfortunately, the very fact that top graduate programs do such a good job of training people to do research and publish means that they often do an inadequate job in other areas of career preparation. You cannot go to graduate school without encountering thorough research training, but you will have to be a resourceful and inquiring student to receive even modest training in other aspects of an academic career. Students recognize this fact. In a recent survey of science graduate students across the United States (Golde and Dore 2001), one of the biggest complaints of Ph.D. students was that they were given insufficient training for teaching and other non-research careers in science. Teaching is routinely seen as a distraction from research. Varying faculty role models will be few and far between in an intense research-oriented graduate program. Pressure from peers, and even from faculty, will often convey the notion that a job at a small state teaching college is for losers that cannot get a "real" job. Bear in mind that as a graduate student, you may find that those around you may not actively encourage you to seek out opportunities to evaluate the full range of post-Ph.D. job options (including those outside academia; Robbins-Roth 2005) and to prepare yourself for the responsibilities you will face in these jobs. Let's consider each of these issues.

One contributing factor to this problem is academic parent-offspring conflict. To some degree, you and your major advisor or postdoc supervisor have similar interests. You want to collaborate in doing good publishable research. However, at this point your interests can begin to diverge. The advisor's research program (and the training of research-oriented students) is best served by having students devote as much time as possible to being in the field or lab doing research. Advisors profit from seeing their students placed at large research-oriented universities. A student seeking broader career experience or training will often come into conflict with advisors or supervisors who are under pressure to do one thing—bring in grant money. Students who wish to attend teaching workshops, write their own small grants, or request teaching assistantships for a couple of semesters will often receive little sympathy or, at worst, be denied any opportunity to undertake this broader career development.

A second constraint on graduate students' ability to explore and train for various career settings is the fact that most graduate programs offer a narrow range of role models. For students whose career goals might take them toward jobs at more teaching-oriented universities, training at a tier I research university poses challenges. These students will have little or no opportunity to gain firsthand experience with the day-to-day realities of a job at a different type of institution. If students believe that their professional needs would be met by a job at a small or midsize college, they will find it far more difficult to make this assessment than will students who want to pursue jobs at first-tier research schools. The latter will have intimate daily experience with faculty mentors who demonstrate the rewards and frustrations of their career choice; the former will have fewer such role models.

The major research university places one last constraint on the ability of students to freely evaluate the range of academic career options. Because the day-to-day activity in research universities is the production and critical evaluation of original research, it is all too common for research productivity not only to determine the types of academic jobs that people will obtain, but also to be used as a metric to measure the inherent value of different academic jobs. Along with their unquestionably valuable research training, graduate students in many programs also absorb the notion that productive researchers get good jobs at "real" universities; less productive researchers settle for positions at teaching schools. Implicit in this culture is the notion that there is a *scala naturae* of universities from the lower teaching colleges to the higher research universities, and only losers are left swimming in the evolutionary backwaters. This is an attitude that students should look beyond and advisors should discourage. Research is research. It is an important part of your career training, but it carries no inherent value above and beyond good teaching or productive service. The value of the career path a grad student or postdoc chooses should be evaluated in light of their individual career goals, and there must be room within our graduate training for students to evaluate objectively the fit between their career goals and the academic jobs available to them.

The take-home message for graduate students and postdocs is that the very thing that makes research institutions so successful—bringing students into intimate working relationships with top scientists in an environment with strong pressure to publish—also hampers those institutions' ability to provide training in other career skills and creates a culture that often provides a distorted view of the job landscape. Good or bad, this is the reality within which most students train for academic jobs in biology. We advise you to understand this dynamic and be sure that you seek out the career experience and training that will serve your needs.

## A Preparation Checklist

So how do you ensure that your graduate and postdoctoral training provides you the best possible preparation for any academic job? This checklist will help.

*1. Think carefully about what career best suits your needs and skills.*
In our experience many people pursue a Ph.D. because they are fascinated by the biology of a certain species or system. However, pursuing your love of biology does not necessarily translate into effective career planning. Academic biologists vary widely in the time they devote to research, writing, teaching, service, and administration; there are non-academic options as well (Robbins-Roth 2005). Early in your graduate work, you must peek up over your binoculars or lab bench and focus on the job setting for which you want to prepare.

*2. Train at a quality research institution, but be sure it meets your own individual needs.*
That Columbia Ph.D. is probably not worth it if you spent all your time fighting with your advisor and peers, lacked committee support,

got little exposure to the broader academic landscape, and hated the New York Yankees. A more conducive professional situation facilitates productive work, positive professional connections, and—let's face it—your own happiness.

*3. Establish a coherent and independent research program.*
At the end of your graduate and postdoc training, you will have to convince a group of faculty that you are an independent research entity (not a cog in your advisor's research machine). Think early about where your research questions are leading, and push to establish signposts of your own independence (side publications, grants, etc.). Be able to describe your research program clearly and in non-technical language.

*4. Expose yourself to as many and diverse role models as possible.*
Your advisor is only one possible role model. Your advisor and committee members, particularly at a major research institution, represent a very limited sample of the types of faculty positions that exist. A minority of the universities in the United States are doctoral-granting institutions. How your advisor at Indiana University spends her day has little to tell you about how a faculty member at Earlham College, just a few hours away, spends his day. Go out of your way to meet faculty from other academic settings. Explore the costs and benefits of each setting. Determine what faculty members do outside of a major research university. This should be as important a goal at professional meetings as meeting faculty in your research area.

*5. Get real teaching experience.*
Many graduate students spend time as teaching assistants for introductory courses, and this is valuable experience. However, to really understand what teaching is like and to get a feel for the role teaching plays in the day-to-day activity of a faculty member, you should make every effort to gain further experience. Consider teaching a section of a large lecture course during the summer. Perhaps a sabbatical replacement is needed for one semester. Even as a teaching assistant, you may be able to move up into upper-level courses with more rigorous content. This experience will give you a reality check for an activity that will occupy a lot more of your time as a faculty member at most schools than it ever does during your graduate career. Many schools will find this experience a real plus in your application.

*6. Do not be influenced by peer pressure regarding the value of different types of jobs.*

There are happy, productive faculty members in all sorts of academic settings. Be sure that your years of training are preparing you for the setting that best meets your needs. Academic training is like a fast-flowing stream that will tend to deposit you at the doorsteps of a major research institution. If that's where you want to be, great! If not, you will have to keep a clear focus on where you want to go (and be a strong swimmer).

# 4

## *Target Your Job Search*

In a behavior known as broadcast spawning, the male eleven-armed starfish (*Coscinasterias muricata*) releases his sperm into the ocean, in the hope that one of his millions of sperm will randomly drift into contact with an egg from a female of the same species, outcompete any other sperm, and so lead to a successful fertilization. At the other end of the spectrum, male snow geese (*Chen caerulescens*) are socially monogamous and form lifelong pair bonds.

The difference between starfish and snow geese provides a useful metaphor for different job-search strategies. When we serve on search committees, we see that some applicants are like starfish, while others are like geese. The starfish applicant sends us the academic version of junk mail. His or her brief, dry, and totally useless cover letter sticks closely to the formula of "Please consider my application for the <insert job title here> position at <insert university here>." Slightly better, but still indicative of the same problem, is the application that shows amusing evidence of its evolutionary history. The cover letter might be addressed to the University of Georgia, include a paragraph about how the applicant is looking forward to working in the Midwest, and state a career goal of working at a small liberal arts school. Both of these applications come from broadcast spawners. Most of you have probably encountered nervous postdocs who have applied for more jobs in one year than the cumulative lifetime total for the authors of this book. Broadcast spawners typically produce applications that suffer from the sort of pitfalls described above. For job searchers employing the broadcast-spawning approach, the job-search strategy can be summed up as "anyplace, anywhere, anytime."

And why not? Many argue that it only makes sense to "toss an application in" to every available search. If lottery tickets only cost the price of a stamp, wouldn't you buy them all the time? We disagree for two reasons. First, as illustrated in the paragraph above, mass-produced applications tend to be superficial applications. These applications do a poor job of explaining why you are a good candidate for the specific job in question. If you do not take the time to link your skills to the job in question, the search committee is unlikely to do it for you. You will not produce your best applications if you are simply sending a generic cover letter and identical materials to every school. Second, and most importantly, any job is not the right job. If you are flooding the job market with applications, you are certainly applying to many universities that are not well suited to your strengths and career goals (unless you are completely indiscriminate or exceptionally gifted). At best, you will later lose interest in these schools and will have wasted your time and the search committee's. At worst, you will accept one of these jobs and become the bane of every department: a disgruntled faculty member, dissatisfied with your professional situation and not really able to serve your students, your colleagues, or your own career.

We can hear the cynic saying that "any job that produces a paycheck is the right one for a starving postdoc." And besides, one's first job can always serve as a stepping-stone to a second, better job. We disagree, but we understand the pressure to take a job—any job—after years of difficult graduate training and postdoctoral work. First of all, why not do the work required to make your first job the one you want? Second, it can be a lot harder than you might think to move to another university after you accept your first job. Once you settle into a job—and begin to face the realities of real-world teaching and service loads—it can be difficult to maintain the research productivity you might need for that dream job (alternatively, a heavy research demand might preclude gaining the broad teaching experience you need to move to that job at the top-flight liberal arts school). As you begin to gain years toward tenure and climb the salary ladder, moving becomes increasingly problematic. Spouses and kids may settle into good careers or schools. We know many faculty members who arrived at a new job with plans to make a quick move to another institution only to settle into permanent dissatisfaction. Of course, we also know many who arrived at their new position with plans to leave, only to realize that they had landed the perfect position. Thus, broadcast spawning is an attractive strategy, but it has its pitfalls. We recommend that you be more discriminating and target your job search carefully.

### Factors to Consider

What factors go into targeting your search only to those positions that will best serve your interests and goals? The previous chapter discussed the many broad decisions you must make during your graduate and postdoctoral training in order to decide what kind of academic position is right for you. Here we consider the specific questions you should ask yourself as you search through the job ads.

Apply for jobs only in those geographic areas that will realistically work for you. How far are you really willing to move from your extended family? Are North Dakota winters or Alabama summers a reality you can put up with? How will your research program fit into different geographic regions? If you are a marine biologist who works on intertidal invertebrates, then a job in Kansas may not serve your needs (or at least may place even greater pressure on you to find grant funding for travel). Be attentive to geographic realities. In our experience, geographic considerations are often a primary—if not *the* primary—determinant of how happy faculty are with their jobs. Do not underestimate the role that location will play in landing the right job. However, keep an open mind. Do not let lack of familiarity lead you to reject regions that may, in fact, offer very satisfying opportunities.

Next, ask yourself if a university with a job opening has a mission consistent with your career goals. University web pages are an invaluable source of information on the mission of the school, as well as the relative roles of teaching and research. Consult these pages *before* you prepare an applica-

tion. In particular, pay attention to the nature of the student body, the relative roles of graduate and undergraduate instruction, research and grant
productivity, course offerings, faculty-to-student ratios, and the background of the current faculty. This information will give you a good feel for
whether the emphasis that the university places on research and teaching
is a good match for the emphasis *you* place on research and teaching. As described in chapter 1, this is an easy call at the extremes. Small colleges will
obviously place a priority on teaching, and an applicant can expect twelve
contact hours per semester. Large doctoral institutions will emphasize
publishing and external grants. It is the many midsize universities between
those extremes that present a real challenge in a job search. There is wide
variation in teaching loads and research emphasis among midsize universities, and this emphasis can change quickly. We cannot emphasize enough
that you should not assume that you know what a university offers based on
its reputation. In particular, many medium-sized regional schools (particularly in the Sun Belt) have grown explosively in recent years, with concomi-
tant growth in research, graduate programs, and so on. That university you
have never heard of may have twenty thousand students and a fine graduate
program. There is no substitute for careful research into a school before
you apply.

What role do you want graduate education to play in your career? Will
you only be happy directing Ph.D. students? Will master's students be an

acceptable substitute, or would you be happy having undergraduates help you with your research? Do not apply for jobs that do not offer the sort of graduate program you require. Again, web pages offer all the information you should need about a prospective school. Some professional societies offer guides to graduate programs in particular fields.

Once you target a school for a potential application, assess whether the department is likely to provide the kind of environment in which you can prosper. Are the department's research and teaching emphases consistent with your own? Will you have potential collaborators, or will you be the only flannel-clad plant ecologist in a department full of molecular geneticists? Will you be able to teach the courses you want? Remember that existing faculty may already teach your favorite course. Is the department's organization conducive to collaboration or interaction? Consider whether you want to be in a full-service biology department or will be happy in a department of genetics or botany.

All of these suggestions seem like a great deal of work just to decide whether or not you want to apply for a job. However, recall that it is all too easy to take a less than optimal job—the "bird in the hand" syndrome—and become mired in an unsatisfactory professional situation. You need only invest a few hours in evaluating the suitability of potential jobs to dramatically enhance the chance that you will become a satisfied and productive professor. At the very least, we hope that our advice will temper the natural tendency of new job seekers to be somewhat undiscriminating.

### Cheating the System

Most people in academia have heard of the "practice interview." This is when candidates apply for jobs they do not really want and accept interviews in order to polish their interview skills. There is much to be gained from this approach—a chance to see how your job talk goes over, an opportunity to discuss your work with colleagues, a few free meals, and a few extra miles on your frequent-flier account. Despite the perks, this is an approach that we do not advocate. You reap all the benefits by gaining experience with the interview process. The department pays the cost of the time and money spent on a candidate who does not want the job, and another applicant pays the cost of getting bumped from an interview slot. Those following the advice in this book would not apply for a job they do not want, but we are assuming that most people are acting ethically and in good faith.

There is little the system can do to prevent such behavior, but we urge job applicants to always deal with prospective employers in good faith. Remember that search committees do talk to each other. If one university suspects that you came in for an interview without real interest in the job, this information can easily find its way to the search committee where you really are interested in a job. If you want to practice, call some of your friends and colleagues together at your home institution.

That said, bear two things in mind. First, when you receive an offer from a department that you are keen to join, having multiple offers can sometimes greatly strengthen your hand during negotiations (see chapter 10). Second, once you are a newly minted assistant professor, your colleagues may tell you that "to be paid what you're worth, you need to get an outside offer." We'll leave that discussion to others.

**5**

## *The Application*

You open the latest issue of *Science* over your morning coffee, and there it is: the perfect job. The Department of Molecular Neuroecology at the Starbucks' Institute for Integrative Studies (part of the Jerry Garcia Center for Big Science) needs an assistant professor conducting funded, cutting-edge research on the interface of neurobiology and landscape ecology. You own that field! As a bonus, the school is located in a stimulating city (yes, they have a vegan restaurant) and recreational opportunities abound. While visions of start-up money, lab space, and skiing on weekends dance in your head, you throw together an application and trust your future to an overworked FedEx employee.

A hard reality waits at the other end of that FedEx delivery. A search committee of four faculty members has received applications from 223 people who all seem to own the field of neuroecology. The members of the search committee—in their spare time between conducting research, giving lectures, working with students, meeting grant deadlines, and maintaining a personal life ("You look familiar, are you one of my kids?")—will have to digest thousands of pages of application materials. Even the most conscientious member of the committee will be hard-pressed to assimilate all the information available (sorry, not all will read every reprint that you sent). Furthermore, each committee member brings his or her own set of preferences, prejudices, and preconceptions to the table. Your application for that dream job will have to negotiate complex and somewhat unpredictable terrain against strong competition.

Many applicants assume that superior training and a history of productivity will be enough to see them through the application process. While these ingredients are necessary, they are often not sufficient. Even the most highly qualified applicant will have significant competition. Furthermore, any candidate's qualifications are open to different interpretations, especially across the range of different academic settings from small teaching colleges to huge research universities—the job seekers' equivalent of a genotype-environment interaction. What one university values (e.g., extensive teaching preparation), another may scorn. What one search committee member sees as a strength ("Wow, twelve years of postdoc experience at Harvard!"), another might see as a potential deal breaker ("Why hasn't this person gotten a job by now? There must be a hidden problem"). We have repeatedly seen our colleagues interpret the qualifications of an applicant in dramatically different ways. Thus, how credentials are presented in a job application can become an important source of variation in a successful job search. Context is just as important as content. A carefully crafted, error-free application can help you catch the serious attention of the search committee and provide a compelling argument for why you are the best candidate, regardless of the perspective or preconceptions of a particular member of the search committee. This chapter will help you craft such an application.

## The Goal of a Good Application

A good job application needs to satisfy one critical goal. It must be informative. However, information must be conveyed *efficiently*, and it must be directly *relevant* to the job for which you are applying. To guide the development of a good application, we suggest that job applicants keep a few important facts in mind.

First, the application is your first chance to make a good impression (or a bad one). A well-organized, well-written application sets a professional tone that can earn your materials that extra bit of attention and that can carry through to other phases of the job search, including a possible interview. Conversely, poorly prepared applications give busy committee members all the excuse they need to shuffle your application off to the rejection pile. The initial screening of job applications is very much an academic version of Clint Eastwood's "Go ahead, make my day." Committee members may look for any excuse to pull the trigger on a rejection. Use your application to make a good first impression.

Second, despite the popular media characterization of idle academics lounging at the public trough, members of search committees do not have a lot of spare time on their hands. A typical member of a search committee must squeeze the review of job applicants between a host of other time commitments. A good reality check for job applicants is to remember that there are real people at the other end of that FedEx delivery who will try to assimilate your application while still delivering a 2 P.M. lecture, meeting a 5 P.M. grant deadline, and attending a stimulating faculty meeting over the lunch hour. Therefore, your application should quickly and efficiently convey information relevant to the job at hand. An effective application is not an encyclopedia of every fact about your professional career that may or may not be relevant to the job in question. There are plenty of job applications that remind us of those ten-page undergraduate term papers with five pages of content. Both make for poor reading. Anything that detracts from the time-efficient flow of information will tend to detract from your application's effectiveness. Help the committee see your potential quickly and easily.

Third, no two members of a search committee will interpret a set of credentials in exactly the same way. The important point for a job applicant is that an application should guide each committee member to the same conclusion (i.e., that you are the best candidate). Just as a politician would never leave the media to interpret political events without some attempt at spin control, an applicant for an academic job should never present credentials without explaining why those credentials make him or her the best candidate. If you cannot give the search committee a detailed and explicit explanation of why you are the best candidate for this job, you are leaving yourself open to the vagaries of decision-making by people you do not even know. Don't assume these truths to be self-evident. Help the committee to see why *your* qualifications match *their* needs.

Fourth, you will quickly discover that crafting a concise job application targeted to a specific job is a time-consuming process. There are no short cuts. However, we believe the results are well worth the time. Fortunately, if you have thought carefully about the kind of job that you want, trained with specific career options in mind, and are applying to a university that meets your needs, then you have already made the important connections between you, your career, and the job in question. In short, you should be excited by this job opportunity and in an excellent position to convey this excitement to the search committee.

Finally, ask some recent hires in your own department if they would be willing to share their application package with you. Whatever its strengths and weaknesses, it will be in any case an example of a successful package. After all, if it weren't, they would not be in your department!

## Components of the Application

Now that you understand what awaits your application, what should you actually stuff into that envelope? Different job ads request different application materials. However, most potential employers will request some or all of the following: a curriculum vitae (CV), a statement of research interests, a statement of teaching interests, reprints, and letters of recommendation. Some will also request transcripts or a teaching portfolio. You may include additional materials (e.g., teaching evaluations in the absence of a portfolio) if they enhance your credentials for the job in question, but use discretion and make sure all supporting materials are concise and easy for the search committee to use. You should also include a cover letter, even if it is not requested in the job advertisement. There are important points to consider regarding each of the components (table 5.1).

## The Cover Letter

The cover letter is arguably the most important component of the application. It is the single document that integrates the diverse components of the application by establishing an overall theme or objective. The cover letter is your spin control; it is the focal point for making your application relevant

**Table 5.1. Components of an academic job application**

| Component | Description |
|---|---|
| Cover letter | A vitally important letter that explicitly describes how your credentials meet the needs of the department as outlined in the job ad. |
| Curriculum vitae | A concise and well-organized listing of your professional experience; includes research, teaching, and service activity. |
| Statement of research interests | A one- to three-page document that describes your research program, with major questions, past experience, future directions, and how the program fits the job for which you are applying. |
| Statement of teaching interests | A one- to three-page document that describes your teaching philosophy, interests, and experience. State what courses you can teach and how you will contribute to teaching in the department to which you are applying. |
| Reprints | A thoughtful selection of your publications reflecting the quality and breadth of your work. If reprints aren't requested, you might include a URL for links to PDF files in the CV. |
| Letters of recommendation | Letters from individuals who can comment directly on your professional skills and on your potential as a faculty member. |
| Teaching portfolio | Only required at teaching-intensive schools. A more detailed statement of your teaching interests and experience, including supporting documents and sample teaching materials. Plays a similar role for teaching that reprints do for research. |
| Transcripts | (Rarely required.) Photocopies of all your undergraduate and graduate course work. Official copies can be sent if needed. |

to the job in question. A well-crafted cover letter leads a reader directly to the strong points of the application. This is an opportunity you should not pass up. Your credentials may speak for themselves, but they speak more effectively when paired with a good cover letter. A CV without a thoughtful cover letter is like the results section of a scientific paper without an introduction and discussion.

Box 5.1 offers an example of what not to do. The letter is your first (and may be your only) chance to impress readers of your application file with your suitability as a job candidate. This first letter simply misses out on the opportunity entirely.

---

**Box 5.1. Example of a poor cover letter**

Dr. Alex Prevezar
Department of Zoology
Big University
Weston, NV 89008
prevezar@bigu.edu

Chair, Ecology Search Committee
Department of Biology
Major State University
Metropolis, NY 00000

Dear Sir or Madam,

Please find enclosed my application for your recently advertised position.

Sincerely,

Alex Prevezar

---

Let's have a look at a better example (box 5.2). Although there is no limit to the specific style that a cover letter might take, the example we provide has the advantages of illustrating the basic design features important to a cover letter and being a real letter that actually landed a job offer. The names have been changed to protect the innocent.

A good cover letter should have the following information:

### 1. The job you are applying for

The letter should explicitly state what job you are applying for. It is not uncommon for departments to be advertising two, sometimes similar, jobs at once. It is surprisingly easy for an application to find its way into the wrong pool. We also believe it is a good idea to provide a bit of context for your application. What is your current position? What attracted you to this job? Why do you want this job? If you already hold a tenure-track (or otherwise permanent) job, this is the place to explain why you are looking to move.

## 2. Your research program

Briefly summarize your research program. You do not need to go into detail; this is what the CV and statement of research interests are for. However, this is the place to explain why your research program is well suited to the department you hope to join. Does it complement and/or add to existing departmental strengths? Is it especially appropriate geographically? Does your work lend itself to undergraduate involvement that the department emphasizes? What skills will you bring to the department? State a convincing case for how your research program meets the needs of the department. You might also describe, in a sentence or two, your research record. How many peer-reviewed publications do you have? Are they in high-quality journals?

## 3. Your teaching experience

Next, provide the same sort of concise assessment of your teaching. What skills do you bring to the table, and how are these relevant to the department? What teaching gaps will you fill in this department? To what existing courses can you contribute?

## 4. Your service or professional activity

Summarize your service or professional activity. For graduate students or new postdocs, this service might be limited. Nevertheless, virtually every applicant should have some service that they can point to as evidence that they will be an engaged and active academic citizen. Did you serve on a graduate student organization? Perhaps you helped with a scientific meeting at your university. Have you chaired sessions at meetings? More seasoned applicants will definitely have university and professional service that should be summarized briefly.

## 5. Your references

Provide the names of people from whom you have requested letters of recommendation. In some cases, a department will ask you for names of references, but will only request letters from a short list of candidates after initial review.

## 6. The closer

Close the cover letter with a concise one-paragraph summary of why you are a strong candidate for this job. Emphasize each of the areas that may be of concern to the search committee: research, teaching, service.

**Box 5.2. Example of a good cover letter**

Dr. Alex Prevezar
Department of Zoology
Big University
Weston, NV 89008
prevezar@bigu.edu

Chair, Ecology Search Committee
Department of Biology
Major State University
Metropolis, NY 00000

Dear Dr. Greene:

Please consider the enclosed application for the Ecology position in your department. I am a vertebrate ecologist seeking a tenure-track position that allows me to pursue excellence in research and teaching. I received my PhD under the direction of Dr. Robert Johnson at Crossroads University, and I am currently a postdoctoral fellow in the lab of Dr. Robin Avis at Big University. I believe I can offer your department a number of specific strengths.

I have developed a funded research program that focuses on the ecology of vertebrates in wetlands. As described in my statement of research interests, I am pursuing two lines of research that are well suited to your department's traditional emphases. First, I am exploring the role of competition in structuring bird communities in freshwater marshes. A major goal of this work is to integrate modeling and field experiments to understand the relationship among water levels, resource abundance, and bird community structure. Second, I am investigating the effects of an introduced fish predator (the Eurasian Perch) on the life history of two species of sunfish. I am employing a series of field experiments to determine how perch are impacting the demography and life history of the economically important sunfish. Furthermore, I hope to use data from my study to help develop effective management plans to reduce the impacts of the invasive Eurasian Perch.

In addition to providing abundant opportunities for student involvement, my work features field experiments, rigorous quantitative analysis, and field techniques such as radio-telemetry and bird banding. Overall, I can offer your department a diverse set of research

*State what position you are applying for.*

*Establish your current situation.*

*Provide a clear abstract of your research program.*

*State clear objectives or goals.*

*Highlight student involvement.*

skills that are a good match for the needs outlined in your advertisement, and would complement existing strengths in your department.

Say why you are a strong candidate.

I also have the experience and commitment necessary to provide quality teaching at both undergraduate and graduate levels. My principal teaching interests are upper-level, field-oriented courses in ecology and vertebrate biology. As a graduate student I taught laboratories for introductory biology, general ecology, and ornithology. Prior to graduation I was given full responsibility for a lecture section of environmental biology. In my current postdoctoral position I teach a graduate-level course on community ecology. My academic and research background also enables me to teach wildlife ecology, conservation biology, behavioral ecology, and evolution. My strong background in statistics (both univariate and multivariate) means that I can comfortably teach (or provide consulting in) biostatistics and experimental design. The enclosed sample of student evaluations reflects the sort of teaching quality that I can offer Major State University.

Provide an abstract of your teaching interests.

Especially for teaching-intensive positions, a list of courses you could teach is a good idea.

As described in my statement of teaching interests, overall, I believe I can make a solid contribution to teaching and advising needs at Major State.

I can also contribute to your department through my active professional and university service. At the professional level, I serve as secretary for the New York Ornithological Society, and I was on the local committee for last year's Regional Ecologists Meeting. At the university level, I served as president of the Graduate Student Organization at Crossroads University. I would continue to be an active academic citizen at Major State.

Highlight important service.

I have requested that letters of recommendation be sent directly to you from Drs. Robin Avis, Curtis Darwin, and Robert Johnson.

Include names of references in your cover letter.

I believe the skills outlined in my C.V. are ideally suited to the needs described in your advertisement. I can offer you (1) the ability to conduct diverse and productive research on the ecology of terrestrial vertebrates, (2) documented teaching skills and a strong commitment to quality teaching, and (3) a willingness to participate actively in the academic community. I hope to have the chance to discuss my qualifications and research interests in person. Thank you for your consideration.

Sum up your strengths at the end of the letter.

Sincerely,

Alex Prevezar

\*

It will be obvious from this description that the cover letter must be customized for each and every application. A generic, all-purpose cover letter does not make a case for why you are the best applicant for this job (it leaves that connection to be made by the search committee), and it does not reflect serious interest on the part of the job applicant. Also be sure that your letter strikes a confident, but not haughty, tone. A self-assured, confident letter will maximize the attention you get from the search committee (especially given that many applications will be accompanied by brief, bland, uninformative letters). Recognize that the letter is a guide or abstract to a "layered" application. The reader gets an overview of your entire application packet in the cover letter. He or she can then delve into the CV and statements of research or teaching to get more details. The reprints or teaching portfolio then provide a third, even more detailed, layer of information.

## Curriculum Vitae

For the search committee, the curriculum vitae is the focal point of their deliberations. It is the one piece of your application that will be handled more than any other. For you the applicant, the CV is the tangible product (and validation!) of long years in training, and it carries an enormous investment of self-esteem. Thus, it is difficult for you to construct a CV with complete objectivity. The temptations are to add, to elaborate, to explain, to reach too far. Resist these temptations and produce a CV that concisely lists and describes your significant professional accomplishments. It is this CV that will best meet your needs and those of the search committee (box 5.3). Good CVs come in many shapes and sizes, but here are the major requirements.

The order in which people list items in a CV is highly variable. The sequence can be varied depending on the nature of the job (an application to a major research university might move publications closer to the beginning). However, in a clearly labeled and concise CV, order should not be a major issue.

### 1. The mundane details

At the top of the first page, provide your name, a current address, voice and fax numbers, and e-mail. It is not a bad idea to include a second "permanent" address (especially if you are in the field or in transition between positions). The same applies to phone numbers—provide a number where you can be sure that, at least during business hours, the caller will find a human being on the other end of the line. While some applicants still describe personal details like marital status, hobbies, birthplace, and so on, we do not believe these are a necessary, or even desirable, part of a good CV. List your personal web page if you have one, but recognize that people thinking about offering you a job might be looking at the website. If you have been meaning to update it, now is a good time. If all your links say "under construction," it's better to include no web page at all.

### 2. Education

List, with minimal elaboration, your educational background. Where and when did you receive your undergraduate degree, M.S. (if applicable), and Ph.D.? You do not need to include the title of your master's thesis or Ph.D. dissertation.

### 3. Professional experience

List, with modest elaboration, your professional experience that is relevant to this job. Many job applicants still seem to evaluate the quality of a CV by its length, not its content. Newer job seekers should resist padding this section with every minor professional experience. Give the search committee a broad picture of your professional training; they do not need to see a month-by-month accounting of your entire graduate career. Typical  entries in this section will be postdocs, academic jobs, teaching assistantships (do not list these on a course-by-course basis), research assistantships, and professional experience outside academia (private labs, environmental consulting, etc.). Consolidate and combine generically similar experience where possible. Those seven jobs over three summers where you banded and censused birds can be nicely summarized in one line item.

### 4. Awards and honors

List any awards that show your abilities to excel. Did you win a teaching award for your work as a TA in biochemistry, or a prize for the best poster at

**Box 5.3. Sample of a curriculum vitae**

### Alex Prevezar
http://www.bigu.edu/prevezar.html

| | |
|---|---|
| Department of Zoology | Office: (222) 555-5653 |
| Big University | Fax: (222) 555-0842 |
| Weston, NV 89008 | Home: (222) 555-5184 |
| (222) 555-5497 | prevezar@bigu.edu |

### Education

| | |
|---|---|
| 2004 | Ph.D.   Department of Biological Sciences, Crossroads University, Clarksdale, MS 34345 (Dr. Robert Johnson, advisor) |
| 1999 | M.Sc.   Department of Biological Sciences, Small University, Fairfax, VA 21341 (Dr. Curtis Darwin, advisor) |
| 1996 | B.S. (honors)   Department of Biology, Ivy University, Cambridge, MA 01020 |

### Professional Experience

| | |
|---|---|
| 2004–Present | *Postdoctoral Researcher: Big University*<br>I am responsible for investigating the effects of competition on community structure in the lab of Dr. Robin Avis, and I coordinate all aspects of a large NSF-funded project (field crew of 5–7 people) during the summer at Douglas Lake Biological Station. |
| 2003–2004 | *Research Fellow: Crossroads University*<br>Held a competitive research fellowship from Crossroads University. Responsible for full-time research toward my dissertation. |
| 2003 | *Instructor: Johnson College*<br>Taught a 5-week summer field course in vertebrate natural history. |
| 1999–2003 | *Teaching Assistant: Crossroads University*<br>Taught introductory labs in environmental biology and upper-level labs in ornithology and comparative anatomy. Responsible for developing introductory lectures, supervising labs, and grading. |
| 1997–1999 | *Teaching Assistant: Small University*<br>Taught introductory labs in environmental biology and general biology. Responsible for developing introductory lectures, supervising labs, and grading. |
| 1996 | *Research Experience for Undergraduates: Ivy University*<br>Served as an independent undergraduate research assistant in the lab of Dr. Cameron Jones. Responsible for PCR work and data analysis. |

### Awards and Honors

| | |
|---|---|
| 2004 | Best Student Paper, Annual Meeting, Ecological Society of America |
| 2003 | Research Fellow, Crossroads University |
| 1999 | Award for Excellence in Teaching, Department of Biological Sciences, Small University |

| 1997 | Award for Excellence in Teaching, Department of Biological Sciences, Small University |
|---|---|

## Publications

**Prevezar, A.,** R. Avis, and E. G. Turgidson. In press. Community patterns in variable environments. Science 34:99–106.

**Prevezar, A.,** and R. Johnson. 2007. Statistical issues in the study of community structure. Ecology 11:104–9.

**Prevezar, A.,** and C. Darwin. 2006. An important study of bird biology. Bird Biology 54:67–71.

Jones, C. D., and **A. Prevezar.** 2004. A molecular phylogeny for minnows. Molecule 45:23–45.

## Grants

**Prevezar, A.** (co-PI, with M. I. Feynman). (Pending) "Nanotechnological methods in avian community ecology." National Science Foundation, $376,000 (2008–2012).

**Prevezar, A.** "Novel approaches to getting your dissertation finished." National Science Foundation, Dissertation Improvement Grant, $20,000 (2003).

**Prevezar, A.** "The analysis of community structure." American Ornithologists' Union, $3,000 (2000).

**Prevezar, A.** "A fundable study of birds." Sigma Xi Grant-in-Aid, $445 (1998).

## Invited Seminars

| 2006 | Department of Biology, University of Michigan |
|---|---|
| 2005 | Department of Biology, Cornell University |
| 2001 | Department of Zoology, Williams College |

## Presentations

| 2005 | Jones, W. D., and **A. Prevezar.** A revision of the minnows. Northeast Fish Society, Boston, Massachusetts. |
|---|---|
| 2005 | **Prevezar, A.,** R. Avis, and E. G. Turgidson. Community structure in birds. American Ornithologists' Union, Santa Barbara, California. |
| 2004 | **Prevezar, A.,** and R. Johnson. A novel statistical approach to analyzing species diversity. Regional Ecological Society, Chicago, Illinois. |

## Major Service

| 2004–Present | Secretary, New York Ornithological Society |
|---|---|
| 2005 | Member, Local Committee, Regional Meeting of Ecologists, Big University |
| 2002–2004 | President, Graduate Student Organization, Crossroads University |

## Professional Organizations and Affiliations

| | |
|---|---|
| American Ornithologists' Union | Ecological Society of America |
| Animal Behaviour Society | Sigma Xi |

a meeting? Have you been elected to Sigma Xi? List those items that would make your grandma proud.

### 5. Publications

List all peer-reviewed publications, in chronological order starting with the most recent. In a separate section, you may include unpublished papers, but only if they are currently in review. Do not include your thesis or dissertation as a publication. We advise against listing any "in prep" papers. While some may encourage you to list these papers, unless they are at least at the "submitted" stage, they are not in the scientific domain. Given that it is easy to list nonexistent papers, review committees discount them, and too many "in prep" manuscripts will simply look bad, especially if they turn up a year later as still "in prep." Use a separate list for unreviewed papers, gray literature, reports, and so on. Do not feel obliged to include all of these other publications unless they really add to the strength of the CV.

### 6. Grants

List the grants that you have been awarded. Include the name of the agency, the duration of the award, and the total amount of funding, in direct costs, over the course of the grant. Include unawarded grants only if they are submitted and in review. For graduate students, even modest internal grants of a few hundred dollars demonstrate good grantsmanship.

### 7. Seminars

List all presentations you have given at professional meetings or conferences or at invited seminars at universities. As your professional experience increases, you may wish to divide this section into contributed presentations (meetings, etc.) and invited seminars (at other universities). This section may get sufficiently long for more experienced job seekers that it is justifiable to include only presentations from the last few years.

WHO? I'M SORRY, DON'T KNOW ANYONE BY THAT NAME.

### 8. References

Although there will be some sensitive cases where listing references is not advisable (see the section on letters of recommendation), most CVs should provide a final page with a standing list (with complete contact information) for three to six references who can comment on your credentials. Be sure that each one has agreed to serve as a reference and that you have chosen these references carefully.

### 9. Miscellanea

In addition to these "core" sections, you may also include, depending on your level of experience and the nature of the job, a list of professional affiliations, your record of service (including the names of any journals or granting agencies for which you have acted as a reviewer), students you have trained, and technical skills. Teaching experience should be included in your list of professional experience. For research-intensive positions, you do not need to devote a lot of space to a list of all the courses for which you have been a TA. However, if you have actually taught courses, this should be included in the CV. Discussions of research interests, teaching philosophy, or abilities should be placed in the research and teaching statement, rather than in the CV.

*

As we have emphasized, make sure your CV is concise and focused on the job in question. Give drafts of your CV to people of various academic backgrounds (teachers, researchers, other students) for comments and editing. With one swipe of a red pen, they can cut away that one-month undergraduate research assistant job to which you are emotionally attached but that is not contributing to the strength of your CV now that you are three years into a postdoc position. Also evaluate the style of your CV. Every member of a search committee has horror stories about CVs whose purpose seemed to be to hide information. Use clear, easy-to-locate headings, keep the number of sections to a minimum, avoid long text entries in favor of concise lists, and stick to the point.

## Statement of Research Interests

The statement of research interests provides a concise but detailed (about one to three pages of text) description of your research program. Remember, the cover letter has briefly summarized your research and how it will fit into the job in question. The statement of research interests is the place to provide the next level of detail (reprints would be the third, most-detailed level). Be sure to address the major conceptual issues that your research deals with, specific approaches that you use (experiments, lab work, field studies, mathematical modeling), and the systems with which you work. Highlight the major results of your work to date. Some of our colleagues find one or two figures of the major results of the applicant's research program useful. Finally, be sure to give a concrete idea of where your program is going. Demonstrate that your ideas have the breadth and depth to keep you busy for at least the next five years of research. Your reader will also want to see that your work is fundable. If you are working in an especially competitive field, you need to make the case that your big plans will not be scooped before you have even had the chance to obtain funding.

Your statement should cite specific publications that act as good road signs along the path that your research program is following. If not specifically requested, you may want to include reprints of two or three publications with your application. This approach will allow the search committee to delve into your program at a broad level (the statement) or at a more detailed level (reprints). Make sure that the research statement (like all parts of the application) is carefully written and designed to be user-friendly.

Finally, you might wish to point out specific opportunities for collaboration or student involvement. This last suggestion will again require you to link your future to the job you are applying for. How will this position serve your research program, and how will your program contribute to the department you hope to join?

We have included samples of two real research statements (boxes 5.4 and 5.5). One is an example of a lengthy, detailed research statement, while the second is an example of a more general and concise statement. Of course, your own research statement will depend on the nature of your work, how far along you are in your career, and the particular university to which you are applying for a job.

## Statement of Teaching Interests and Philosophy

The statement of teaching interests is the teaching equivalent of your research statement. Give a one- to three-page statement that describes your teaching experience, your teaching philosophy or approach, and where you see your teaching "program" headed in the future. It is unfortunate that many applicants for academic positions do not seem to view their teaching as a coherent program of activity (as they readily do for research). But a career in teaching is a coherent program of activity in which you teach certain

## Box 5.4. Sample of a detailed research statement

**Research Interests**
**Population Genomics of Transcriptional Regulation**

Selective, demographic, and random processes all determine the frequency of alleles in a population and differences between species. One of the major goals of population genetics has been to uncover which of these processes is acting in natural populations through a combination of directed empirical studies and theoretical models that provide expectations under a variety of conditions. Until recently, most of the work in this field has involved single-locus or limited multiple-locus studies and models. Now that genomic-scale data are available, we are in need of new genomic-scale approaches. Within the next year, a population of whole genomes will be sequenced; the approach that population genetics takes now may determine how soon these data become informative and what information they give us.

My research program is focused on developing the empirical, computational, and statistical tools necessary to study variation in whole genomes in an evolutionary context. More specifically, I am studying the evolution of cis-regulatory sequences—the DNA necessary for directing the time, level, and place of transcription of protein-coding genes. The sequencing of whole genomes and a growing number of studies into cis-regulatory variation have shown that the effects of natural selection reach far beyond the base pairs that fall between the start and stop codons. Through a combination of directed empirical studies, new computational techniques, and improved statistical tools, my goal is to contribute to our understanding of the role cis-regulatory variation plays in evolution.

**Ongoing Research**

My current research can be divided into three main areas: computational approaches to studying populations of genomes, the development of models for statistical inference of natural selection, and empirical studies of cis-regulatory variation in humans. Below I briefly describe my prior research in these areas and my future plans for research.

*Computational Biology*

The availability of whole genome sequences means that we can examine natural selection not only on thousands of individual genes, but also at the level of the genome itself. Because cells must regulate the transcription of suites of genes expressed together but located throughout the genome, and because transcription factors may control hundreds of target genes expressed at different times and places, selection for improved transcriptional efficiency may act throughout the whole genome.

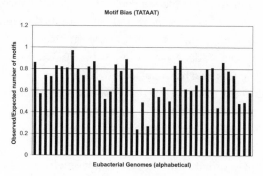

One promising target for genome-wide selection is that of transcription-factor binding sites. These short sequences (generally 6–10 base pairs) make up motifs that appear frequently throughout a genome. If new binding sites are created frequently, this may allow novel transcriptional patterns to evolve. At the same time, however, these new sites may introduce a large degree of noise into the efficient functioning of a cell. I hypothesized that the binding of transcription factors to spurious binding sites—the correct sequence of nucleotides in inappropriate genomic locations—could drive natural selection to eliminate binding-site motifs from a genome. In a study of 52 whole genomes in Eubacteria and Archaea (see above figure), I developed novel tools to demonstrate that spurious binding sites appear less frequently than expected under a random model in every genome but one (Hahn et al. 2003). It appears that both functional and nonfunctional sequences are constrained to avoid mutating to binding-site motifs. In addition, I developed a model of binding-site evolution that allows an estimate of the strength of selection against binding sites. Selection intensity appears to be weak, similar to that of codon bias.

I am now extending these studies of motif bias to eukaryotic genomes, where heterochromatin, gene-rich and gene-poor regions, recombination, and multiply represented binding sites greatly complicate both the effects and detection of natural selection. These studies will greatly expand our understanding of natural selection, patterns of variation, and the regulation of transcription.

## Statistical Inference

To understand how selection, mutation, and drift can affect within-species variation, we need explicit population genetic models. With these models in hand, we can draw statistical inferences about the forces that act on genes and genomes. A major component of my research, therefore, aims to develop population models appropriate for the analysis of both individual genes and whole genomes.

In concert with my work on binding-site motifs described above, I have developed statistical models of motif bias among populations of genomes. Using analytical and simulation techniques, I have shown (Hahn and Rausher, submitted) that individuals will vary in the number of any particular nucleotide motif simply because of single nucleotide polymorphisms, even in the absence of selection acting on motifs. By adding natural selection to these analyses, I find that selection reduces both the mean number of motifs in the population and the amount of variation among individuals. Based on these theoretical results, I have been able to create a likelihood ratio test to determine whether natural selection acts on individual binding-site motifs in actual genomes.

For the analysis of individual genes, use of coalescent genealogies has made for more efficient and more precise statistical inference. However, both demographic and various selective mechanisms will cause significant deviations from the neutral-equilibrium model. In order to distinguish demographic from selective effects, and among different selective mechanisms, I developed an improved statistical test based on the coalescent (Hahn et al. 2002). This test has revealed hidden instances of natural selection on coding and cis-regulatory sequences not detected before (Hahn, in press), and has provided a powerful tool for other researchers (e.g., Schaeffer 2002; Mes 2004; McDaniel and Shaw, in press).

## Regulatory Variation

In order to understand the role of natural selection, it is also important to study the individual mutations with phenotypic effects that are visible to selection. In an ongoing study of functionally characterized cis-regulatory polymorphisms in humans and the other primates, we have been collecting sequence data among human populations and from chimpanzee, bonobo, gorilla, orangutan, and baboon. This work focuses on binding-site polymorphisms within a regulatory region that have been shown in biochemical studies to significantly affect transcription levels in humans. Working with Dr. David Goldstein at University College London and Dr. Gregory Wray at Duke University, we continue to find evidence that selection has acted on cis-regulatory mutations acting in specific human populations (Rockman et al., in press) and also on humans in the past (Hahn et al., in prep). The enormous amount of functional cis-regulatory polymorphism in humans implies that there is not one static regulatory network seen by natural selection, but rather a population of networks. This work will have important implications for both the evolution of modern humans and for the evolution of transcription factor: DNA interactions.

## Future Work

In addition to pursuing the research topics outlined above, I also have a number of additional studies planned that will complement this research. One major study aims to look at variation in genome sequences among individuals of the mosquito, *Anopheles gambiae*, an important vector of human disease. With the availability of a population of genomes, we will be able to examine the patterns of variation and selection at every gene and across the genome. I am collaborating with a group of researchers from the European Molecular Biology Laboratory, the European Bioinformatics Institute, and the University of Notre Dame's primary population geneticist on a project to sequence multiple genomes of *A. gambiae*. On a finer scale, I have begun a project with mosquito researchers here at UC Davis to look at polymorphism and divergence in a number of *A. gambiae* loci thought to be important for disease resistance.

To fully understand the enormous amount of data generated by genome sequencing and functional genomics projects, we need an appropriate evolutionary and population genetic framework. In the next few decades, the field of biology will no longer be data-limited; we will be limited only by the analytical tools that are available. My research aims to both create and facilitate the use of these evolutionary and population genetic tools. I believe that a vibrant research laboratory comes through having a mix of people with different backgrounds and expertise. I look forward to building a lab that uses empirical, statistical, and computational approaches to understand the evolution of genetic diversity.

courses using a specific approach to yield desired outcomes. Teaching is not just a pesky activity that breaks up your research schedule. Unless you are an exceptional researcher, you will directly impact far more individuals through your teaching than through your research. (One of us calculates that he has taught over three thousand students so far in his career, far more than the number of people who have read some—most? all?—of his technical papers.)

Specific points to convey in your teaching statement include the courses you have taught in the past, the courses that you are able or want to teach, how your teaching is suited for the department you hope to join, and what kinds of approaches you will emphasize. Do you like to work from current events? Do you emphasize the process of doing science and current literature? Are field trips a major part of your teaching? How do you interact with students? Are you comfortable at all levels of teaching (non-majors, introductory, upper level, graduate)? What is your yardstick for success as students leave your courses? Although styles may vary, this statement gives a good idea of the sort of information you are trying to convey. We provide

an example here from a senior academic, which we hope will give those of you just starting out a better perspective on the sorts of things you might consider (box 5.6).

### Letters of Recommendation

Your CV includes a list of references, and you have stated who your references are in the cover letter. Job advertisements will request either that you have letters sent directly to the search committee, or that references may be asked by the search committee to send letters at a later date. In either case, be sure to follow appropriate etiquette. Give your letter writers plenty of

**Box 5.5. Sample statement of research interests**

My primary research interest is the evolutionary ecology of vertebrates. In particular, I am interested in understanding the causes and consequences of (1) morphological and behavioral variation within populations and (2) patterns of habitat selection and use. The existence of ecological differences among individuals from a single population challenges our understanding of how variation is maintained within populations. However, it also offers a unique opportunity to quantify the factors that play a role in the origin and maintenance of ecological diversity. It is this opportunity that forms the basis for one of my principal research interests. How and why organisms select particular habitats, especially against the background of increasing human manipulation of the environment, is the other.

One of my lines of research grows out of my postdoctoral work with Dr. Ellen Ketterson on the effects of hormones on the breeding biology of a socially monogamous bird, the dark-eyed junco. Hormones such as testosterone play a critical role in controlling alternative reproductive strategies in birds. By manipulating testosterone levels (with silastic implants), it is possible to experimentally create phenotypic variation within a population and quantify its consequences. In juncos, testosterone-boosted males shift effort away from the care of current offspring (reduced feeding of young) and into behaviors that may increase mating success (singing, extra-pair fertilizations). My research seeks to discover how testosterone mediates such fundamental shifts in reproductive behavior. Using radiotelemetry, I have shown that much of the change in male behavior is mediated by testosterone's effect on spatial activity (*Animal Behaviour* 47:1445–55).

Manipulation of hormones in concert with radiotelemetry offers a unique opportunity to explore the ecological and evolutionary consequences of hormonal variation in free-ranging vertebrates. My long-term goal is to continue to exploit this system (in collaboration with Dr. Ketterson) to further understand how testosterone affects a male's response to conflicting environmental cues (e.g., fertile females, young in the nest, intrusions by other males, etc.). Through the summer of 1994, I will be using radiotelemetry to quantify how testosterone affects the trade-off a male must make between guarding a fertile mate and seeking additional copulations off-territory.

Another line of research concerning variation within populations deals with morphological and behavioral strategies associated with avian migration. This research has included analyses of intraspecific variation in wing shape (e.g., *Auk* 109:235–41) and migratory timing (e.g., *Condor* 92:54–61). In collaboration with Dr. Frank Moore, my studies of avian migration are now exploring patterns of habitat selection and their relationship to successful stopover along the migration route. In spring and fall, individual migrants moving along the Gulf and Atlantic coasts are faced with selecting an appropriate habitat in which to stop over and replenish fat reserves. This decision must be made under varying levels of experience (age), energetic stress, competition, and predation. Furthermore, stopover sites are increasingly subject to human development. How these factors interact to determine habitat selection is unknown despite the fact that migrants often stop over in discrete habitat patches (e.g., cheniers, barrier islands, coastal woodlands) that provide tractable units for the experimental study of migration.

My goal is to exploit this system in long-term studies of habitat selection and stopover success. Currently, I am using graphical models to predict how migrants of varying energetic status (i.e., fat levels) should respond to various combinations of habitat and competitors (these results have been presented at meetings and manuscripts are in preparation). My interest in diminishing coastal plain habitats is also extending into bird community dynamics in pine savannas. My goal is to understand how management of this increasingly rare habitat will affect the composition and dynamics of its bird communities; I have three years of funding from the U.S. Fish & Wildlife Service for this work.

In addition to my primary research interests, I have extended my work into other systems (e.g., habitat relationships between snakes and anoles in Puerto Rico; *J. Herpetology* 24:151–57) and into conceptual issues of evolutionary biology (species concepts and speciation; *Systematic Zoology* 38:116–25). My field-research experience includes work with birds, reptiles, and mammals. I also strive to apply novel quantitative analyses and field techniques (e.g., correspondence analysis, hormone implants, radiotelemetry) in my studies.

**Box 5.6. Sample statement of teaching interests and experience**

I have broad teaching interests and abilities, but my principal skills are in the area of organismal biology. My strengths as a teacher include a commitment to quality instruction, proven teaching abilities at the introductory and upper levels, diverse teaching experience, and the ability to establish a rapport with students of all levels. I enjoy teaching, and I welcome the opportunity to interact with students, both in and out of class. In all my teaching assignments, my goal is to provide students with a course that is up to the minute in content, impeccably organized, and taught with clarity. I think my success in meeting this goal is reflected in the enclosed sample of student evaluations. My ultimate goal as a teacher is to be a positive influence in the professional development of my students.

I have taught at the introductory level on many occasions, including courses for majors and nonmajors. My responsibilities have included oversight of graduate teaching assistants, development of new laboratories, and teaching of up to 350 students. In these introductory courses, my philosophy is to establish a link between the subject matter and the day-to-day experiences of the students. I strive to provide students with the background they will need to understand contemporary biological issues and to make informed decisions on current events (particularly environmental issues). As you will see in the enclosed evaluations, my students have responded enthusiastically to this approach. Despite my active research program and my interests in upper-level courses, I enjoy teaching at the introductory level and find it rewarding to interact with beginning students.

My primary teaching interests are upper-level, field-oriented courses in ecology, evolution, and behavior. These interests, combined with extensive teaching and research experience, give me strong and diverse teaching abilities. My experience includes teaching assignments in general ecology, community and ecosystem ecology, animal behavior, evolution, biometry, conservation biology, ornithology, herpetology, and vertebrate biology. Through these courses I have firsthand experience teaching the biology of all four classes of terrestrial vertebrates, as well as principles of ecology, evolution, and behavior. I have also taught a number of graduate-level seminars on current research issues in evolutionary biology. Finally, I have experience teaching gifted undergraduates (animal biology within the honors curriculum at Indiana University) and teaching field-research methodology (sampling methods and experimental design at Oklahoma's biological station).

In upper-level courses, both graduate and advanced undergraduate, my philosophy is to provide solid content while emphasizing the process of doing science. My courses stress familiarity with current research problems, direct experience with experimental methods (through independent student projects), and observation of organisms in the field. By introducing students to current questions in a particular discipline, giving them the opportunity to plan and carry out field experiments, and providing the chance to see organisms in natural habitats, I try to give students hands-on experience in biology. My goal is to produce students that not only are knowledgeable in a field, but who can also apply that knowledge to solving real research and management problems. My own research on the ecology, evolution, and behavior of vertebrates gives me an excellent background from which to teach upper-level courses. Furthermore, I enjoy working with advanced students, and I believe they enjoy my teaching (I enclose some evaluations from recent courses).

In addition to the courses described above, my academic and research background enables me to teach widely within the fields of ecology, evolution, and behavior. I can readily teach courses in evolution, behavioral ecology, experimental design, and population biology. Overall, my diverse teaching experience and documented teaching skills allow me to teach effectively at both the introductory and upper levels.

Courses taught: introductory biology for majors, introductory biology for nonmajors, environmental biology, herpetology, ornithology, general ecology, community and ecosystem ecology, evolution and ecology, terrestrial sampling methods, conservation biology, biometry, experimental design, vertebrate biology, natural history of the terrestrial vertebrates, honors animal biology, field studies of the Gulf Coast.

advance notice (a few weeks if possible; at least as much as the application deadline will allow). Every advisor can recall—with clenched teeth—sitting down for a day of productive work only to have a student or postdoc rush in with a request for an important letter of recommendation that must be mailed by tomorrow. "Well-organized," "timely," and "conscientious" are not terms that readily find their way into letters written under these circumstances. Remember, most graduate and postdoc advisors write a steady stream of letters for their students and employees. While every advisor is happy to support the career of his or her academic offspring, this task is time-consuming and usually taken for granted. Incidentally, this is another good reason to apply only for those jobs that you are serious about considering; do not waste the time and goodwill of your references by applying to every job in the back of *Science*. Otherwise, your references may tire of sending all those letters and may be inclined to write bland, generic letters that will not be effective.

Provide your letter writers with a written request that includes the deadline, a copy of the job advertisement, and a pre-addressed envelope for their convenience. Also consider providing a copy of your application letter or a brief description of the highlights of your application (what strengths will your application stress?). This will help your reference to add comments that

might be especially appropriate for a particular position, but do not expect them to customize your application for a specific job. That is your job in the cover letter. Finally, send your reference a polite reminder/follow-up as the deadline draws near (one to two weeks). As long as you are polite (Post-its on the office door saying "I need that letter today" are a no-no) and have given the writer plenty of advance warning, most letter writers will appreciate that you are following up on your earlier request. You will not need a reminder if your reference has already confirmed that the letter was sent.

Who should write these letters that for all intents and purposes will determine your future? In principle, references are easy to select. Choose people who (1) can speak directly to your abilities and potential (advisors, supervisors, employers), (2) have good standing in the academic field in which you hope to work, and (3) can write a strong letter for you.

For most readers, this will be a straightforward issue. You probably are well-known and appreciated by your academic advisor, your postdoc supervisor, and perhaps another member of your committee. You will likely have no further concerns.

However, an unfortunate (and mercifully small) minority of people have run into conflicts at one time or another with a supervisor. Or perhaps there is a conflict between one of your references and someone in the department to which you are applying. Here are a few things to keep in mind:

First, the search committee is going to expect certain people to be among your references, including postdoc advisors and your graduate advisor. If you do not include a letter from your grad school or postdoc advisor, some suspicions will be aroused. What do you do if you believe one of these "expected" letter writers would not write you a strong letter? You cannot afford uncertainty in this important area. There is nothing wrong with asking a letter writer for a candid assessment of the letter they will write for you. In fact, we encourage more communication between applicants and their letter writers. Most references should be willing to characterize the tone of the letters they are producing for you. A word of warning, though: Do not ask about your letter unless you are prepared to hear the truth. Instead of getting mad and pouting, consider that a writer is doing you a significant favor by telling you that they cannot write more than a lukewarm letter. You do not want to be in the situation we observed recently where a letter writer said he "really could not remember" the person for whom he was writing a letter! If an advisor or supervisor cannot write a strong letter, then you will need to omit a letter from an expected source. Of course, it is possible that this person declined to write a letter in the first place, a preemptive statement that they could not write a good letter. In either situation, we believe your best bet is to address briefly why an "expected" letter is missing. The place to do this is in the cover letter section that lists your letter writers. A carefully crafted sentence can usually downplay this issue sufficiently so that it will not compromise your application: "Although I enjoyed a productive working relationship with my graduate advisor, Dr. Jones, our personal differences are such that my postdoctoral mentors Drs. Smith, Doe, and Hawk can more effectively speak to my professional and personal characteristics."

Second, be an informed consumer. Know the personality landscape surrounding a job opening. Who is in the department you hope to join? Can you get wind of who some of the other applicants are? If you can identify potential conflicts of interest or personality conflicts up front, you are in a better position to choose appropriate references or discuss possible problems with your letter writers.

Third, if you have sensitive situations that may affect a job application (a spouse who is also on the job market, health problems, etc.), you will want to address these issues on your own terms and at the time of your choice. Be *certain* that your letter writers know how you want to approach these topics. We have often read letters of recommendation that broach touchy personal matters not mentioned in the applicant's cover letter. Even more problematic, letter writers often do not cast these issues in the most positive light. A frequent approach is for the writer to identify a sensitive subject such as a job-hunting spouse as a "problem" that they are confident the applicant can overcome. Maybe so, but search committees do not readily go looking for hiring problems when there are two hundred applicants in the pool. Be sure to discuss with your references any off-limit topics or any special requests about how they handle sensitive issues. Similarly, if there is any information that you would like for the committee to know of, but not from you, you might talk about this with one of your references.

## Teaching Portfolio

A teaching portfolio describes and documents your teaching goals and accomplishments. Compared to a statement of teaching interests, a portfolio would contain a much more detailed statement of goals and approaches, and it would include substantial supporting materials like class handouts, course web page samples, tests, student evaluations, and so on. Edgerton et al. (1993) and Seldin (2004) provide good descriptions of teaching portfolios, so we will restrict our discussion to the role the portfolio plays in a job application.

The statement of teaching interests and the teaching portfolio have a relationship that is similar to that of the research statement and reprints. The statement of teaching interests provides a brief overview of your teaching program (we urge you to consider your teaching to be a coherent program of activity), while the teaching portfolio lays out the detailed supporting evidence for your teaching creativity and productivity (the "scholarship of teaching" in current lingo). Just as you select reprints to include in an application based on their quality, creativity, and programmatic relevance, the portfolio should document creative, quality teaching that is relevant to the job for which you are applying.

Like many other portions of a good application, a teaching portfolio should be targeted to a specific audience and will take time to prepare. This is not what most graduate students or postdocs want to hear, especially given that few job advertisements specifically request a portfolio, and virtually all advisors are cracking the whip for more research.

The more cynical would say that teaching portfolios are for losers who cannot get a "real" job and must settle for the consolation prize of a job at—dare we say it?—a teaching college. As should be clear by now, we disagree with this sort of reasoning, and we offer a couple of thoughts on the use of portfolios in job applications.

The teaching portfolio will be most important in applications to colleges where teaching is a priority (actually a priority, as opposed to colleges where teaching priorities do not get much beyond marketing materials). At schools where teaching is emphasized, your teaching portfolio will be as important or more important than your research activity. If you are applying to a school like this, your teaching portfolio should be detailed and carefully crafted. Articulate why you teach and what outcomes you want.

Also, keep in mind that a good teaching portfolio will continue to pay off in the future. Once you find yourself in a tenure-track job (even at a research institution), you will be evaluated annually for progress toward tenure. Increasingly, teaching portfolios are the way (or one of the ways) that faculty demonstrate their teaching productivity. Getting in the habit of thinking about your teaching program and documenting this program in a portfolio will give you a head start in the tenure-track environment.

### Final Considerations

Make sure all components of your application support the theme established in the cover letter. Each and every part of the application should be "on message." The more coherent and targeted your application, the more likely it is to get the serious attention of the search committee.

Does your application reflect knowledge of the university? Failure to know the most basic facts about the university to which you are applying can lead to sometimes comical, often damaging, gaffes. Applicants to Georgia Southern—a 16,000-student comprehensive university with strong emphasis on both research and teaching—have at different times pointed out how much they would like to join the faculty at a small teaching college, a small liberal arts school, or a university in the Atlanta metropolitan area (that three-hour commute from our campus here in south Georgia is going to be tough). While these sorts of mistakes are not fatal, they do create an impression of hasty, unprofessional work.

You should proofread carefully. It is stunning how frequently we see proofreading errors in applications. If a candidate's application for a highly competitive and potentially permanent job has spelling and grammatical errors, can this person be expected to produce careful and accurate work on a daily basis? Hasty preparation can also lead to the sorts of mistakes mentioned above. It is hard to believe that we would need to stress the importance of a complete, professional, and error-free application (isn't this the same lecture we just gave to our undergraduates concerning term papers?). Nevertheless, we have been dumbfounded by the frequency of applications with poor organization, serious errors of omission, and repeated typographical errors. Errors are unprofessional and reflect a lack of care in preparing the application; these errors can only lessen your chances of being the successful candidate.

Finally, do not send out an application until you try it out on some colleagues. You are not likely to send out a manuscript for publication until you have sought comments from people whose opinion you respect, and the manuscript would never be published without peer review. Surely your job materials should be afforded the same careful consideration. Let friends and colleagues read through your basic application. Swallow your pride and ask them for frank comments and suggestions.

# 6

## *Preparing for the Interview*

All those hours spent selecting appropriate jobs and crafting a good application finally pay off when the phone rings. On the other end of the line is the search chair. Chances are he or she is saying that you are on a short list of four to eight candidates; are you still interested in the position? Most candidates at this point ponder how many milliseconds they should pause—you do not want to sound desperate!—before saying yes. If all goes well, this call will be followed fairly quickly by an invitation to come for an interview or, in some cases, this may be preceded by a phone interview. The former is obviously the most important, but both types of interview require you to make similar sorts of preparation. In fact, our experience suggests that failure to prepare properly for the interview is one of the major reasons a candidate fails to get a job offer. Do not put yourself in the position of the job candidate who exclaimed, "If I had known you did so much research here, I would have planned a more research-oriented seminar!" You may know your seminar verbatim and have a charming personality, but you still need to do a lot of homework before you get on that plane to interview for your dream job.

If you are preparing for an interview, you have made it through the most difficult steps in the job search—at least the steps with the longest odds. Depending on how many people applied for the job, your odds have improved from 300:1 to something closer to 3:1. You are now down to a head-to-head competition, and your performance is completely under your control. In particular, the planning you do ahead of time will often decide the outcome, much like the athlete or musician who visualizes an entire performance before she has even begun. Good preparation will allow you to develop themes and talking points that will ensure a strong interview. A disturbingly large number of candidates take themselves out of serious consideration because of a faux pas that ultimately traces back to how they prepared, or failed to prepare, for the interview. The purpose of this chapter is to guide you through the necessary preparation. Keep in mind that this preparation serves the dual purpose of improving your interview performance and providing some of the raw data on which you will make a decision should you receive a job offer. We consider the job seminar, which will be the central focus of your performance, in chapter 8.

## Where to Find Information

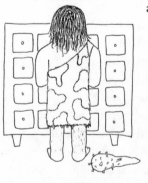

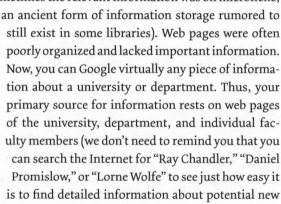

Times have changed since the three of us interviewed and obtained our academic jobs. As late as the early 1990s, the only way to learn something about some universities was to search through the card catalog in your library (sometimes the relevant information was on microfiche, an ancient form of information storage rumored to still exist in some libraries). Web pages were often poorly organized and lacked important information. Now, you can Google virtually any piece of information about a university or department. Thus, your primary source for information rests on web pages of the university, department, and individual faculty members (we don't need to remind you that you can search the Internet for "Ray Chandler," "Daniel Promislow," or "Lorne Wolfe" to see just how easy it is to find detailed information about potential new colleagues). The good news is that you can do your homework with relative efficiency without leaving the office.

Wireless hubs and Internet cafés will take you far in your search for information, but carbon-based life-forms can be extremely helpful too. Contact friends and colleagues who have some knowledge about the school or department. As you may be learning, the field of biology is relatively small and you probably don't have to go too far to find someone who can offer some firsthand information. Someone in your current department is bound to know someone associated with your interview school. Perhaps the best information you can get, however, is directly from someone in the specific department. Many potential candidates think that contacting the search committee is taboo. In fact, nothing could be further from the truth. By the time you have been chosen to come for an interview, the committee *wants* you to do well and they will be more than willing to answer almost any question (but this is not the right time to ask about salary!). The reality is that this search committee has spent hours poring over application packets, and they are pulling for "their" short list to perform well. After all, the search committee does not look very impressive to their departmental colleagues if all the candidates bomb the interview! However, if you are going to contact the search committee, it is best to have a specific question or two prepared. It is best to contact the chair of the committee, as he or she is likely privy to a bit more information than the others and is supposed to be the spokesperson for the department. Bear in mind that while contact is fine, if you call every day, you risk annoying someone at the other end of the phone.

## What Information Do You Need?

The list of information that you might gather about a school is endless. What is it you are looking for? The easy answer is to gather any information that you believe will help you to evaluate the school and to prepare for the visit. But we want to help you be efficient in your sleuthing, so here are a few tips on things to look out for.

### The University

With apologies to Thales, your first maxim should be "Know thy university!" A major focus of this book so far has been that it is in your interest to evaluate honestly what kind of school is ideal for you. If you are actually going to apply for that cell biology position at Coffee College by the Sea, make sure that CCS has the traits that appeal to you. How big is the university? Is it growing or changing in major ways? If so, what are the targets or strategic plans for the university in the next few years? What selling points does the university use (comprehensive research, teaching quality, student-centered)? What are the academic and demographic characteristics of the student body? Is the university building strengths in your own area of interest?

This may seem like pretty arcane information for a job interview, but the present and future situation of the university is critical to your long-term success and happiness in an academic position. If you are looking for a regional university that emphasizes teaching, you do not want to find out after accepting a job that the university's strategic plan calls for massive expansion of research over the next few years. Learning this information also makes it clear that you are a candidate that takes this position seriously, and it puts you in a position to ask better questions regarding your future employer.

## The Department

During your interview you should remind yourself that if you are the successful candidate, you probably will be spending over two thousand hours a year in this building surrounded by these people. Thus, facts related to the department are all-important for your performance in the interview, and for your ultimate decision. Here are the pieces of information that you must be familiar with *before* arriving.

### What Kind of Department Is It?

Some schools have a unified biology department while others have separate units for zoology, cell biology, microbiology, genetics, and so forth. This organization will influence the sorts of interactions you will have on a daily basis as you will likely interact much more with colleagues in your own building than with those on the other side of campus. If you are a person who asks research questions that cut across conceptual areas or phylogenetic groups, you may prefer to have a more heterogeneous pool of colleagues who work on plants, animals, fungi, and protists in the same department. On the other hand, you may prefer to be in a more specialized department where you can join an existing center of strength.

### What Is the Composition of the Faculty in the Department?

If the position is at a small liberal arts college, it is possible that the department will have only a handful of faculty members. On the other hand, a school with a large biology department or a series of divided departments may have from twenty to well over a hundred faculty in the life sciences. In addition to the size of the department, you should also consider the balance of senior versus junior faculty. Departments vary in the degree of faculty turnover and the frequency of new hires. If you are applying to a school that has not had any openings in recent times, this means you could be the only assistant professor in the department. These senior faculty may be good mentors, but you may miss the energy, optimism, and built-in support group that comes with junior colleagues. Ideally, there will be a good balance of faculty at different stages in their careers.

### What Are the Teaching Needs and Expectations of the Department?

Find out what courses are being taught. Are there any of these that you could also teach? The guy who has taught Introductory Microbiology for thirty years may be quite happy to cede the course to a young Turk, or he may jealously guard his turf. Are there any new courses in your area of expertise that you could create? Also, find out the regular teaching load of the faculty and how many courses they teach each year. If you are evaluating teaching loads, it is often critical to know not only how many hours you will teach, but also how many different courses you will be responsible for.

### What Are the Specific Research Needs of the Department?

Does your work fill a void? Are you bringing something in particular that you see as a need in that department? Perhaps you have mathematical or statistical prowess, experience with an exciting new methodology, or expertise in a specific area that the department currently lacks. Alternatively, you may look at yourself as someone whose skills match those that already exist in the department, allowing you to help the department build on existing strengths.

*What Is the Relative Emphasis on Graduate versus Undergraduate Training?*

This is one of the fundamental differences between the different categories of the Carnegie classification scheme. As we described in chapter 1, this will have an immense impact on how you spend your day and the amount of research productivity you can expect out of yourself.

## The Faculty

> *"There is nothing more inexcusable in these days of the Web than not knowing a little bit about the people in the department you are interviewing in."*

During your interview you will meet with a large number of faculty. They will have questions for you. However, they also expect that you will show curiosity about them and their work. Use departmental and personal web pages to learn about your potential colleagues. You do not need to know the names of kids and pets, but at least be able to show familiarity with major research interests, courses taught, and departmental activity (who is the grad program director, for example). Be prepared to be a lively and engaged partner in conversation.

Some years ago, a friend on a job search at a large midwestern university was being taken to a faculty interview by the head of the department. When she announced the name of the next faculty member he would be seeing, he responded enthusiastically, "Oh, I remember a paper of his on heavy metal effects on photosynthesis!"

The department head was clearly impressed: "My, you've done your homework. He hasn't published a paper in sixteen years."

Our well-prepared friend got that job.

## A Final Consideration

There is one final consideration as you get your nerves in check and prepare for the interview. A job interview is a physically and mentally grueling event. You will be kept up for long hours—from an early breakfast to a late social event—and you will be subjected to intense interactions throughout the day. You will answer scores of questions from deans, chairs, senior faculty, junior faculty, and students. Some topics will be covered repeatedly as you move from faculty member to faculty member, but you need to respond with enthusiasm each time. The only way you will negotiate this marathon successfully is to be excited by this job opportunity and to go into the interview with a positive attitude. Think like the well-prepared athlete or musician. As you learn about the research interests of the members of the department prior to your visit, imagine some of the conversations that you might have with students and faculty. You will be more relaxed during your interview, and you and your hosts will all have a more enjoyable experience.

# 7

# *The Interview*

The long-sought interview has finally arrived. All that planning and preparation has paid off, and you now have a reasonable shot at a job offer. In fact, at this point your odds are pretty good. In our experience, out of the three or four candidates who interview, at least one or two will make significant mistakes or give a tepid performance. A good performance during the interview can quickly move you to the head of the class. The most important single event is the seminar, and we consider that topic in the next chapter. Here we discuss what you can expect during the interview process and how you can maximize your performance.

## Getting There

The first thing you need to do is get to your destination in one piece. You need to arrive with the critical materials for your interview and be prepared to meet the first member of the search committee. There are some pitfalls at this stage that you can easily avoid. First, be prepared for a delayed flight. If you are flying from coast to coast during snow season and have a choice of a transfer in Minneapolis or Salt Lake City, choose the airport less likely to be shut down on account of a winter storm. Be sure you are carrying the phone numbers and e-mails of your contacts. If at all possible, notify them of flight delays before they drive long distances to the airport to pick you up. Keep the interviewing department informed about your progress. By the way, once you arrive at your destination, turn off your cell phone; do not take phone calls during the entire interview process.

> *"I was about to start a conversation in my office with a candidate for a full professor position in our department. His cell phone rang, he answered the call, and spoke for twenty minutes, using up almost all the time I had been allotted for our interview. When I told my colleagues this story, they all agreed that this was really arrogant. He didn't get the job."*

A second common problem is that while you head to Texas for your interview, your suitcase gets an all-expenses-paid trip to Kansas. Lost luggage can wreak havoc on an interview if you have put all your eggs in one basket. If at all possible, only take carry-on luggage. If you do check a suitcase, never pack the storage medium holding your talk in checked luggage, and be sure to bring enough clothing and toiletries on the plane to get through  one day. Yes, this could happen to you; in our experience, lost or delayed luggage is a fairly common occurrence. Airline security continues to evolve, but make every effort to carry onto the plane everything you need for the first day. Finally, be aware that the interview will begin the second you are picked up at the airport or hotel. Have in mind some topics of conversation for this initial ride into town or to the campus. Try to establish enthusiasm and show some knowledge of the school and area right from the beginning.

## General Philosophy

OK, assuming you arrived on time and with your belongings intact, what now? The first thing to do is to have fun and maintain a positive attitude. While the interview process is stressful and the intensity of interactions exhausting, with the right attitude you can actually have a wonderful experience. After all, academia is an environment you have decided you enjoy. So enjoy the chance to meet new people and to tell them about the interesting research and teaching that you do. Above all, be yourself. None of our advice should be construed as advocating that you play a role during the interview. Rather use our advice to accentuate a positive presentation of your true self.

One complication about being yourself is that very little about this whole scenario may be familiar to you. How often have you been wined and dined, or at the center of attention? How many invited seminars have you presented to a room full of strangers?

### Dress for Success

> *"There's nothing worse than not taking the whole thing seriously. We had folks come to interviews in jeans and a fleece jacket."*

How you dress and how many clothes to bring depend on the duration of the interview. Dress professionally, but not so formally that you call attention to your dress (you are not applying for a job at a Fortune 500 company).

Clothing should be conservative, comfortable, and not provocative. At most places the interview will be at least two to three days. It is acceptable, and probably a good idea, to dress a bit less formally for the second day (think business casual). You will not only feel more comfortable but will let your colleagues see how your phenotype will likely appear on a day-to-day basis. However, still dress respectfully (no jeans, sandals, or T-shirts). It is also appropriate to dress a bit more casually for informal social events, but again do not get carried away (we recall a male candidate who attended a social in a colorful polyester shirt open to the lower chest, complete with heavy gold chain). Of course, if you are being taken on a field trip or hike with some faculty or graduate students, make sure to bring the proper attire for this outing. A final note on your attire is to dress appropriately for the geographic location of the interview. If you are a postdoc at the University of Miami, you probably need to take into account that in February the weather for your interview at Harvard University will be a tad more inclement. While easy access to online weather forecasts might make this seem rather obvious, one of us made the rookie mistake of not packing a jacket for a "springtime" interview in the state of Indiana (where it was twenty degrees Fahrenheit on arrival). He had to borrow one. On the other hand, there was the case of a candidate who came to interview here at Georgia Southern in early spring and wore only a wool suit. Suffice to say, with temperatures in the eighties, this person almost denatured.

## The Process

Before we go into details about the interview, it is important to understand what goes into deciding who receives the final job offer. Knowing the process will likely impact how you conduct yourself during the interview. After all of the candidates visit, there will be a general faculty meeting at which the performance of you and your competitors will be assessed. Depending on the school, the search committee may or may not bring a specific recommendation to the meeting. The faculty will then vote individually on the committee's recommendation or on the candidate of their choice. While in some cases this vote is not binding, the head of the department will typically support the choice of the faculty. Graduate student input is sometimes an important part of this process too. Given all this, no single person will be responsible for the decision. It is up to you to impress as many people as possible during your brief visit.

## One-on-One Meetings

You probably have heard the saying that you only get one chance to make a good first impression. That is what an interview is: a series of first impressions. It is critical, as we already mentioned, to realize that in most departments everyone has a vote. As a consequence, each first impression is important indeed! To make matters more interesting, each category of person you will meet will be evaluating you differently. Here's what to expect when you talk face-to-face with individual faculty, graduate students, the department chair, and the dean.

## Faculty

As someone who has been in the postgraduate environment for many years and likely in as many as four different schools, you probably know by now that not all professors are the same. In your basic biology department, you can feel pretty confident that you will spend time with the following individuals: (1) the scintillating conversationalist (with luck, the most common phenotype), (2) the interrogator, (3) the bore, (4) the weirdo, (5) the gossip, and (6) the confidant (this is the one who might tell you what life is really like at State U). In addition to their various predilections, some will be more enthusiastic than others.s Some may be active researchers while others focus primarily on teaching and service and have not published a paper since before you were born. In all cases, their opinion matters, and a large part of their opinion will be based on how they evaluate you as a future colleague, friend, or collaborator.

> "The most common gaffe I have seen at the interview stage is not knowing anything about the research interests of people in the group where you are interviewing. I hate candidates coming into my office and asking, 'What do you do?' Take some time and actually look over people's research if you really want a job somewhere."

Do not try to put on a separate show for each meeting. Be consistent by being yourself. However, you should have familiarized yourself with the research or professional interests of each faculty member. Inquire about their work and their current activities. Show that you are engaged and interested in the department. Your interviewers have had plenty of time to evaluate your work, and you can use this as a chance to find out more about theirs (and maybe even start some potential future collaborations). This is where all your preparation from chapter 6 comes into play. Armed with a little information about their background, you should be able to quickly find solid talking points with any faculty member.

Be prepared to discuss your work in greater detail, including your thoughts about what you plan to do in the future. While you will have some opportunity to do this during your seminar, faculty who work in areas related to your own might use this opportunity to probe a little more. They may want to know how much you know about your subject, how creative you are, how realistic, and whether you will make a good mentor for graduate students. At the same time, you will likely be faced with the challenge of carrying on a scientific conversation with someone whose expertise has little or no overlap with your own. Consider this an opportunity to learn something new, and rise to the challenge of finding connections between apparently disparate work.

> "Being uninterested and/or uninformed about people's research is the worst mistake. Don't say things like 'I'm a rat physiologist. Why do they have me talking to you, a pollination expert?'"

In addition to asking about research and teaching, do not hesitate to pose more wide-ranging questions. These questions can provide valuable information, and they reflect interest and curiosity on your part. What do they think about the department? What would they like to see in the person hired for this position? Do they consider the faculty in the department to be interactive? Are there a lot of collaborations among faculty? Do faculty share equipment? (This is particularly important in smaller schools.) How do they like living in this particular town/city? Is the administration supportive of faculty? How is the dean? How do they find working with the contracts and grants office? Is the office a positive force or an impediment when submitting grants? Does the campus offer good support for teaching? A good strategy while interviewing is to ask the same question of different faculty members. Just as in your research, you know that a large sample size is more representative than $N = 1$. Too often when we have asked candidates, "So, do you have any questions for us?" they respond with, "No, I have asked all my questions already." We guarantee you that this answer loses votes. Always have one more question you can ask.

One caution is that as you query various faculty, you will occasionally stumble across dissatisfaction or ugly department politics. While you certainly want to know about such situations, do not be drawn into taking sides or expressing opinions on department issues that you know little about. In more factious departments, some of your interviewers may ask loaded questions to assess which faction you are likely to join. It is not always easy to see these questions for what they are, but just follow the general rule of not expressing strong opinions on issues like department governance, tenure rules (though this is certainly a legitimate area of inquiry), and so forth.

## Department Head/Chair

For obvious reasons, you do not want to mess up this meeting. You will find that the chair is pretty smooth at directing the conversation. The boss knows what he or she wants to find out. In addition to the questions mentioned above, there are some questions that you should direct specifically to the chair. What is the strategic plan for the department? (Normal faculty have no clue what this means, but administrators have visions and strategic plans, and they love to be asked about them.) Does he or she envision more hires in the near future? What are the expectations for tenure in this department?

At many schools, even though you will likely have met the chair earlier in your visit, this is typically the last person you will spend time with. This exit interview is the time for you to make sure you ask all of your remaining questions. It is also the time when you will probably be asked how you feel about your visit. In other words, after spending a day or more, would you be interested in the job if it were offered to you? Financial issues are likely to be brought up at this time. The chair may inform you of the starting salary and will probably be interested in what sorts of things you need for setting up your lab. We cover the details of these negotiations later (chapter 10). Concerning your start-up, we recommend you come prepared to the interview with a typed list. In fact, two typed lists. One is "My Ultimate Wish List" that contains everything you would like. The second one is "My Minimal List of What I Need to Function in This Department." Having the lists already prepared makes you seem

> MY MINIMAL LIST OF WHAT I NEED TO FUNCTION IN THIS DEPARTMENT:
> COFFEE

organized and professional and saves you from having to think of a bunch of items on the spot. As you compile these lists, we recommend that you speak to advisors and other faculty (especially friends who have recently started jobs) so you will have appropriate lists.

## Students

At most interviews the candidates get to spend some time with a group of graduate students or undergraduates. Do not blow this off as an inconsequential part of the interview. They will be evaluating you for your qualities as mentor, advisor, advocate, and, most importantly, teacher (whether it be in the classroom or the lab). Furthermore, students will often provide some of the most "spin-free" information regarding a department and its workings. While faculty might not be totally forthcoming with answers to your questions, the grad students tend to be brutally honest. Depending on your age and experience, they may see you as someone who is just a few years out of grad school, so you will be able to relate on a personal level to the students. Use this as an opportunity to find out what they think about the department. Here are some questions that you can come armed with. How is the level of financial support for students? Are the faculty supportive of the graduate students, or is it an "us versus them" situation? Do the faculty get along, or are there factions? Is the chair effective? Is there a high-quality seminar series in the department? What do they think of the quality of the graduate students? Why are they at this school? What are *they* looking for in this particular hire? While the grad students may not have a vote, you can bet that either formally or informally the faculty will be interested in your performance and the students' impressions of you.

### Dean or Other Administrators

Meeting the dean seems like a frightening part of the interview. In reality, you can usually relax with this one. The dean is usually there to cover administrative details and not to evaluate your qualities as a faculty member. And boy oh boy, if academics like to talk, deans take the cake. However, if you are asked if you have any questions or if there is a seventeen-second silence, you can pull out the old "What are your goals for the department or college?"

### Questions versus Answers

Often at the end of an interview, the interviewer will ask if you have any questions. Always have some questions up your sleeve; this is where your preparation comes in. Good questions reflect interest and good preparation. Of course, be genuine—if you're acting, they'll figure it out.

## Other Possible Interview Events

### Phone Interviews

Smaller schools with fewer available funds may ask you to carry out a phone interview before or even in lieu of a formal visit. If this is the case, keep a few things in mind. First, determine who will be on the other end of the phone. Just as you would with a visit, find out as much as you can about the people who will be interviewing you. You might even want to print out photos of each person, so you can more easily keep track of who is asking questions during the interview. Second, set a specific time for the call, so you are prepared with pen, notepad, "talking points" that you hope to cover during the interview, a list of questions, and so on. If you pick up the phone just as you have stepped out of the shower, you might not have a chance to be fully prepared.

### The Roundtable

"*Probably the most memorable positive interview was a candidate who oozed enthusiasm and ideas at the roundtable grilling that is a feature of our interview process. When asked if the candidate had any research projects that she could start immediately on arrival that would be suitable for graduate students, she laughed and said, 'Do I have any?! Sure. I have so many, I don't know where to start!' and then laid out four or five really excellent, diverse hypothesis-driven projects that were significant and easy to understand.*"

Some schools have an event where some or all faculty are invited to a session where you are asked a series of scripted questions. This allows all the candidates to be evaluated somewhat objectively based on how you answered this set of questions. This is a very important event for you, and we have seen people win or lose the job offer based on their performance here. It is sometimes reminiscent of a Roman amphitheater complete with lions and slaves. The good news is that you can prepare for this. The questions at roundtable events are fairly predictable. Box 7.1 provides examples of the sorts of questions you can expect to hear. Universities will vary in their emphasis, but at all roundtables you can expect to cover research and teaching. Generally, expect to talk about your research and teaching goals, how

**Box 7.1. Examples of questions that might be asked during a roundtable event**

Why do you want to work at this university?

Describe your research program.

What will be the topic of your first major grant proposal?

What other sources will your funding come from?

What are your laboratory and facilities needs?

Where do you see your lab or research program in five or ten years?

How will you measure your success as a scientist?

What sort of projects do you have in mind for students?

What is your approach to mentoring undergraduates or graduate students?

What are your plans for publishing?

In what journals do you plan to submit your research?

How did you get interested in biology?

How will you fit into our existing departmental strengths in . . . ?

What new strengths or skills will you bring to our department?

What courses can you teach?

What courses would you prefer to teach in our department?

Tell us about your teaching philosophy.

How does your research inform your teaching?

What is your approach to teaching in a large lecture setting?

What is your approach to teaching small upper-level courses?

What academic accomplishment are you proudest of?

How do you handle conflicts with students or faculty?

How would you see yourself fitting into a department such as ours?

you approach teaching, where you will seek grant funding, what attracted you to the department, and so on. Keep your answers on point and do not ramble. A brief and organized answer is far better than a Faulknerian one.

*"I was on a committee where, when asked about plans for future research, the candidate said, 'I don't currently have any idea what I would do next,' and then motioned like he was shooting himself in the foot, and actually said, 'I know I really screwed that one up.'"*

Distribute your attention around the room as much as you would in a seminar. Try to establish eye contact and connect with everyone in the room. Finally, in case this is not completely obvious to you by now—be prepared to ask questions of your interviewers. Our experience at roundtables is that many people, when asked if they have questions for us, simply say no. Having most of the faculty in one room where they can respond to questions about the department, its functions, and its direction is an opportunity that should not be passed up.

### One Last Final Piece of Advice

Throughout your interview you will be imbibing coffee, water, soda, iced tea, and so on. From your Intro Biology days, you probably recall that fluids are extracted in the kidneys and stored in the bladder until expelled. It is the rare search committee that, while putting together a candidate's schedule, actually pencils in time to visit the loo. Within the bounds of reason (and necessity), never pass up an opportunity to take a  bathroom break. These moments will come all too rarely. They give you an opportunity not only to empty your bladder (especially important if they have been plying you with lots of tea or coffee), but also to have a minute or two of quiet solitude.

# 8

# *The Seminar*

So far, we have talked about the many factors that go into a successful job application and interview. However, there is no getting around the fact that the seminar is the single most important event during your interview. You will probably be nervous, and your performance in this sixty-minute time slot will, to a large degree, determine your fate in this particular job search. For many (perhaps most) faculty, and certainly for most other members of the department, your seminar is the only chance they will have to see you during your visit. They will be a critical audience, evaluating your dress, your science, and your potential as a teacher. Despite the job seminar's obvious importance, the three of us have countless times found ourselves dumbfounded at the fundamental mistakes many speakers make. Do not underestimate the consequences of a dumb mistake! While the seminar is for the most part under your control, it is a complex performance. We believe that with some preparation beforehand, you will be able to perform in stellar fashion and win the job. In this chapter we provide some suggestions on all aspects of the seminar process.

## The Purpose of the Job Seminar

A job seminar is different in both content and objectives from the myriad of talks you have seen as a graduate student and postdoc. It is longer than a fifteen-minute conference presentation and is not the same as an invited departmental research seminar. The goal of most research seminars is to tell a group of like-minded scientists about your research program. This is part of a job seminar, but there are additional realities. The first point to realize is that not everyone in your audience will be judging you on the same criteria. Many will be interested solely in your research skills. Other will be making judgments about your teaching ability. Still others will be looking for hints about your personality and potential collegiality. Thus, you should realize that you are not giving a "normal" research seminar of the sort you would give at a professional meeting. Your research must be solid and well presented, but do not lose sight of the broader significance of your talk.

In addition to a research talk, some departments will also request that you give a lecture to students in an introductory course. This talk is somewhat less complex than a traditional research seminar because you only have to wear one hat (demonstrate good teaching). However, do not be fooled into thinking this is a slam-dunk. The purpose of a teaching seminar is broader than just conveying information clearly to students. The faculty members in your audience will be assessing your pedagogy. In other words, they are not just interested in whether you explained the Krebs cycle correctly. They are evaluating your teaching style and philosophy and comparing this to current best practices. Thus, the teaching seminar can be a rather complex performance as well. You must engage a classroom full of students you have never met, and you must do this in a rather self-conscious way (how am I approaching the task of teaching these students?).

## Preparing the Seminar

As with so many aspects of the job hunt, a good seminar begins with solid preparation before you leave home. Begin by recognizing that the seminar is the one part of the interview process over which you have total control, assuming no acts of God occur while you are standing in front of the audience (e.g., your PowerPoint presentation freezes, the projector bulb blows, the fire alarm goes off). You should have absolutely no doubts about your ability to give a tight and polished seminar. You should know to the nanosecond the duration of your talk. This means that you have given plenty of practice talks to a live audience (faculty, grad students, and even undergrads). If you can't find anyone, talk to a mirror or to your cat— they will respond with equal enthusiasm. We cannot stress enough how important it is to thoroughly prepare for this event. If you just add a few slides to that fifteen-minute conference talk the day before you leave for the interview, you are in trouble. We suggest that you concentrate on three areas in your preparation.

## Learn from Others

In the weeks prior to your job seminar, reflect on your own experiences as a member of a seminar or lecture audience. While you have surely seen some spectacular talks, you have also been forced to endure some duds. In fact, thinking back on past job seminars, we often can remember the embarrassing ones even more easily than the real winners. Close your eyes and recall some of those nightmare hours you spent in bad seminars during the past few years. Remember wishing that you were sitting in a dentist's chair having your third molar drilled sans anesthetic rather than listening to this talk? OK, then think of why you felt that way. What specifically were the problems? Were the problems related to content or presentation? Were there annoying behavioral traits? Now do the same thing for the great talks. What was it about them that made them positive experiences? Draw upon these past experiences and learn from someone else's performance.

Needless to say, this is best done for a substantial period of time prior to the interview. As with so much else in this process, remember to always look far ahead in your academic career. As a grad student and postdoc, use every seminar or lecture you attend as a training ground to think about how you hope to perform in a similar situation. Pay close attention to style and content whenever you get the chance to attend a job seminar in your own department. Keep a mental list (or an actual notebook) of good seminar or lecturing techniques, particularly effective slide design, ways to encourage audience participation, and so on. When it comes to seminars or lectures, you do not have to reinvent the wheel. Build your own effective style from those that have gone before you.

### Content of Your Seminar

Your first heavy lifting will be to decide on the content of your research and/or teaching seminar. The overall content of your research seminar will be dictated by the specifics laid out in the job ad to which you responded. If you have multiple research projects or if you work across disciplines or taxonomic groups, be sure that you do not stray too far from the job description. Assuming you are on target with your topic, you must next decide whether to talk about a single research project or multiple distinct projects. If you have recently completed your dissertation, the decision may be a relatively easy and obvious one. However, if you have been a postdoc or research associate for a while, you might have two or more potential projects on which you could speak. If you speak on one research project, you will gain the advantage of being able to treat a single coherent project in detail. However, excessive detail can lose audience members who are not familiar with your specific research area, and it can be difficult to place a single project into the broader framework of your long-term research program.

Alternatively, if you develop a seminar around two or three distinct projects, you can highlight the breadth of your skills and can nicely illustrate a multi-themed research program. Be careful here, as you run the risk of presenting a disjointed collection of topics or coming across as a dilettante. In our opinion, the detailed, single-topic talk probably works best at major research universities where you will be expected to specialize in funded, high-intensity research. A broader, multi-project approach might be appreciated more in smaller, less specialized departments. That said, we have seen some fabulous seminars where a job candidate spoke about empirical work for the first half of the seminar and then about a theoretical project for the second half, for example.

Whichever approach you take, it is critical that you begin your talk with a broad introduction that has two goals in mind. First, take the time to provide some background for your work. Why is this an interesting biological question? A surprising number of job candidates begin their talks at a level (complete with discipline-dependent jargon) that leaves only specialists able to fully appreciate why this work is interesting or relevant. Even at the most high-powered of research institutions where top-flight research is the name of the game, all members of your audience will appreciate a few minutes devoted to setting the context and broad significance of your work. You are also being evaluated as a teacher. If you can't take a few minutes to explain to an audience that includes herpetologists or parasitologists why your work with the 4EBP1 protein is critically important, this doesn't bode well for your ability to make the Krebs cycle seem relevant to undergraduates.

Second, you must place the work you are discussing into the context of your overall research program. What are the broad conceptual ideas that led you to this project? Why do you find your research interesting and important? What do the results discussed in your seminar tell us about what motivated your work in the first place and where your research is headed in the future?

The concept of the research program is important. It is meant to convey that your research is part of an ongoing scientific process, not just an isolated project. Setting your work into the context of a research program shows that you are addressing broad conceptual problems. Your program has a past and a future.

At the end of your talk, revisit the broad themes that you established at the beginning. Be sure to tell the audience what biological conclusions they can draw from your work and where the field needs to go next. This provides a nice segue into where your research program is headed in the future. Consider spending five or ten minutes on the range of problems that you plan to explore during your first few years as a tenure-track faculty member. This part of your seminar will also give you the opportunity to relate your future research and teaching plans to the specific department where you are interviewing.

One of the more challenging aspects of this seminar is determining the appropriate level of detail you should use to discuss your work. No forty-five-minute seminar provides enough time to talk about *every* aspect of your work. You will likely need to gloss over some details of conceptual background, methodology, statistical analysis, supporting projects, or experiments to validate methods, and so on. Those ten months you spent carefully backcrossing strains to ensure that your mutant lines were fully co-isogenic may have seemed important at the time, but your audience doesn't need to know the details. The decision about what details to provide is particularly important in a job seminar. If you give excessive detail, you run the risk of losing your audience in a forest of discipline-specific minutiae. Too little detail and people will think your research is superficial or unsophisticated. We provide two guidelines to help you evaluate the level of detail in your talk. First, like a trail of bread crumbs, the entire talk must have a clear path or narrative that any biologist can follow. You can take time to pursue the most detailed and arcane of issues as long as your listener knows why he/she has arrived at this topic and how the outcome of this discussion will relate to broad questions being addressed. Establish a strong narrative in your seminar. Second, at some point in your seminar, you must get your hands dirty and show your audience how you do your science. You want to make the talk user-friendly, but not so superficial that you never discuss molecular or mathematical models, show detailed statistical results, or discuss knotty alternative interpretations. The nature of your seminar will also depend on your audience. If you are a microbiologist giving a talk in a microbiology department, you obviously do not need to cover the difference between Gram-negative and Gram-positive bacteria in the introduction to your work on the ways in which different types of bacteria elicit different patterns of cytokine production in their hosts. If, however, you need to convey the same results to a genetics department or to a general biology department, you will obviously need to come with different sets of assumptions about what your audience will or will not understand. Your ability to set the right tone for everyone in the audience will go far toward showing them your potential as a scientist, communicator, and future colleague.

The question of detail is also important for the teaching seminar. Although departments requiring a teaching talk will often specify the topic (e.g., photosynthesis, Mendelian genetics), it will be up to you how you approach the subject and in what detail. Our advice in the previous paragraph applies here as well. Establish for students a strong theme in the lecture. Tell them why this is important, and provide a road map of where you are going. Just like a job seminar audience, students in a lecture must know how any given piece of the lecture is related to the ultimate take-home message. However, be sure to include significant detail in your lecture. The faculty evaluating you will want to see that you can make complex or substantive material accessible to the students.

One last point about your seminar: You will need to convey to your audience that you have a research program that is independent of your grad or postdoc advisors. Many speakers do not make an adequate effort to establish where the advisor's research program stops and where their own program begins. This can be a difficult part of the seminar. Obviously you do not want the viewer to leave your talk thinking that your research advisor provided the questions and methods that you discussed and you acted as a technician. The best solution to this challenge is to think about your research program early in your research career (chapter 2) and to actively seek out opportunities to establish your independence as a researcher. During the seminar, your independence will come out in your ability not only to describe your past research, but also in your plans for the future, and the way in which you handle questions.

### Slides

Once you settle on the content of your job seminar, you will need to translate this decision into PowerPoint slides. The most elegant and compelling seminar in the world can still go down in flames because of bad slides. On the other hand, good slides can make even the most difficult material seem like child's play. As with content, we recommend that you pay close attention to good and bad slides in the talks of other people. Here are some guidelines that we find useful.

### How Many?

No one ever complained about a seminar or lecture being too short. The duration of your talk will largely be a function of the number of slides you use. It is difficult to provide a precise upper value because that depends on the information contained in each one. For example, talks by ecologists or field biologists may have numerous slides of field locations or organisms that one can move through rapidly. Nonetheless, a rough estimate is that you will spend one minute per slide. If you average much more than a minute per slide, audience attention can tend to lag as you begin to spend large stretches of time talking while the visual aids are static. However, if you take only a few seconds per slide, you will be moving too fast to convey information effectively. When you factor in the time during your presentation when you will be talking without slides (e.g., introduction, summary), thirty to forty-five slides would likely be a comfortable number. Personally, we recommend a conservative estimate of the number of slides you will need. Never include slides that you think you "might" get to if time permits. When a speaker begins skipping slides near the end of a talk, it reflects poor planning, and it weakens the presentation. It is a good idea, however, to have slides in reserve that you think might help answer likely questions from the audience. These would be covered only if a particular question arises at the end. It is a thoughtful and well-prepared speaker who can field a question and say, "That's a great question, and I have an extra slide that more fully illustrates that analysis."

### How Much Information Per Slide?

Slides are visual aids. They should aid your ability to convey digestible bits of information in a brief period of time. You will do this by using some combination of text, figures, drawings, photos, and tables.

For text, limit yourself to about five lines per slide. This seems brief, but as text proliferates it can become distracting, with audience members reading ahead or losing interest in long text passages. Furthermore, in our experience a lengthy text passage on a slide almost always repeats what the speaker is saying anyway. While your text slides can emphasize take-home messages, there is no need for you to read what is on the slide. Use text sparingly, in digestible chunks, and in ways that are not redundant with the spoken information.

It is easier to understand a quantitative pattern in a graph than to read and interpret numbers in a table, so try to convert your message into graphs whenever possible. Each graph should have a title and labeled axes (and in your talk make sure you read these to the audience!). Try to make graphs that are simple enough to convey one to two points efficiently. If the graph is inherently detailed (e.g., sequence data, phylogenies), use techniques (see below) to isolate the important pieces of information. For tables, make sure you include the *minimum* amount of information necessary to make your point; it takes time for an audience to absorb rows and columns of numbers. Never present a slide that requires you to say, "This table has more information on it than you need, so please only look at column 23." If the audience only needs to look at one column, then only include one column in the slide. Never apologize for a slide—replace or omit it. In general, remember that no matter what you say, the attention of audience members will wander to all parts of your graphs and tables. Thus, don't include superfluous information.

Which font should you use? Although tradition is that one should use sans serif fonts (without those little curly tips on the letters) in a presentation, we feel it is acceptable to use anything that you and the folks who saw your practice talks think is fine. However, beware the purist in the audience. According to one font expert we know, the best way to convey information is to use a sans serif font for titles or headings and a serif font for the body of the text. In all slides, make your letters large enough to be read easily in poor light from the back of the room by people who may be a bit nearsighted.

### Title and Acknowledgments

Although the three of us do not agree on everything, we do agree that title and acknowledgment slides are not critical to a job seminar (and of course are not needed in a teaching seminar). The title is not necessary because there have likely been notices around the building for a week already. Also, the host just mentioned the title when you were introduced. Acknowledgment slides rarely convey any meaningful information for your audience. Of course you had an advisor, a slew of undergraduate helpers, some funding, supportive parents, and a loving dog. But no one in the audience likely knows their names, so there is no reason to show them. If you have collaborators, you will mention them as you describe the work. If you feel uncomfortable with the notion of not showing this slide, consider showing the list at the end and simply saying, "I would like to thank the many people whom I worked with in the lab and the funding agencies who supported this work." Do not read your way through a long acknowledgment.

As for the introductory slide, you might start with a picture that conveys the general concepts of your talk, one that you could show to an audience of non-scientists at the local Kiwanis Club. On the formal side of things, consider beginning your talk with a slide that outlines the major components of the talk. This "road map" slide can help you place the talk in the context of your research program, and it can reappear at major transitions to provide a reference point for the audience (or help them catch up if they have been reading the newspaper). A useful concluding slide is one that presents your future research plans. Your audience will like to hear how these plans will fit into the department and what sort of potential your work will have for involving students and attracting funding.

### Wow! Look What PowerPoint Can Do!

PowerPoint in the hands of a biologist is like a Ferrari in the hands of a teenager—both have ample power to get the user in trouble. All those colors, patterns, and effects that seem so neat on the screen in your office are largely distractions in a job seminar. Look at any textbook, journal, or newspaper—what colors are the text and background? Bingo—black on white provides maximum contrast. This color combination is also optimal for any sort of room. In dark rooms, the white background provides some light for your audience to take notes and makes it harder for them to fall asleep or surreptitiously grade exams. In a light room, the contrast makes it more likely that your audience will see the slides. If you are strongly in favor of light lettering on a dark background, just be sure to choose simple, contrasting colors, such as white or light yellow text on a very dark blue background.

Just because you have the technology to place rotating sienna letters on a flashing aquamarine background doesn't mean you should. If you insist on using colors, keep them simple and limit the number of colors. Refrain from using red and green together, as some percentage of the population is color-blind. Colors perform best when they are used judiciously as a highlight or accent to focus attention in a slide. For example, if you have inherently complex graphs (like sequence data or phylogenies), use color to highlight where readers should focus attention. We have also seen colors used effectively to convey statistical significance.

Do not give into the temptation to use patterned backgrounds. We have never seen a presentation that was enhanced—and we have seen many that were weakened—by the use of patterned backgrounds. Also avoid PowerPoint's many special effects unless they can clearly enhance complex visual information (e.g., have a complex structure appear sequentially). We are unanimous in our opinion that we would much rather see plain black on white than colorful and patterned slides that leave you squinting to make out the data, and we don't need to see data appear on-screen to the screech of tires.

## Transport

The corner store can offer quick replacements if you forget your toothbrush or underwear. However, they will be fresh out of CDs burned with a copy of your talk on lipid-mediated regulation of hepatic monoacylglycerol acyltransferase. Thus, you cannot afford to forget or to lose your presentation. The safest approach is to bring your storage medium with you onto the plane (do not pack your talk in checked luggage). It is also a good idea to e-mail a copy of your talk ahead to whoever is coordinating your visit. At some universities you may be able to upload your talk to a departmental server, or you may have it available as a download from your home server. Although most computers will play nice with each other these days, consult with the person coordinating your visit to be sure that platforms are compatible. The easiest thing to do is to bring your own laptop, and any connectors that may be necessary to ensure cross-platform communication. If you are using an on-site computer, make sure before you leave home that your Mac-designed presentation works on a PC or vice versa.

## Presentation and Practice

### To Read or Not to Read

Best not to read. Oral presentations in science are typically not read, and we have all seen wunderkinds give flawless and engaging seminars without so much as a scrap of paper for notes. This is a worthy goal for your job seminar. However, remember that the purpose of your seminar is to convey information clearly and effectively. If you need notes to do this, then use notes, but keep the notes telegraphic. Although talking with notes is a bit more formal and less interactive, reading a text will be seen as very awkward by the audience unless the room is full of people just back from the annual Modern Language Association conference. If you read a text, the audience of scientists may be so focused on the strange apparition of a seminar read from text that they'll forget to pay attention to the science. You run the risk of creating a barrier between you and your audience.

Generally, scientists can avoid text because we have slides as a persistent visual mnemonic device. But be careful. You should not stand with your back to the audience as you use the slides as a crutch, just because you're determined to avoid notes. It is surprisingly common to see speakers stumble and struggle through job seminars as if they had never given the talk before.

You can avoid all this trouble with a simple solution. Practice. But if you need to read, then read. The bottom line is that you should do what it takes to give a clear and well-organized talk.

## Time

Most universities run on hourly schedules, so you probably will have an approximately sixty-minute time slot for your talk. This does not mean you will talk for sixty minutes. Some of those sixty minutes will be taken up by the introduction, a potential late start, and your question-answer session at the end. Many students and faculty will need ten minutes to get to their next class. When preparing your talk, we suggest that you plan on talking for a *maximum* of forty-five minutes. You might want to contact your host ahead of time to find out how long the seminar slot is, and how much time the department typically likes to leave at the end for questions. It is crucial that you leave a sufficient amount of time for questions because this is one of the key events of the interview. In order to stay on time, do not stray from your script. Speakers often ad-lib during the early phase of the talk, but this will only pinch you for time at the end and detract from your performance. During your practice talks, give yourself some benchmark times and know where you should be at each minute so you will be able to speed up if necessary. There is nothing worse than getting forty-five minutes into a talk and hearing, "Ok, uh, now for the next part of my talk." You do not want the critical last few minutes of your talk to be given hastily or to be interrupted by people starting to file out toward their next class.

## Behavior

The pressure of a job interview, public speaking, and presenting your science to a critical audience all combine to bring out a remarkable array of behavioral quirks. The list of possibilities is long and amusing, but this is a serious matter because an audience is more likely to listen intently to your seminar if you do not distract them with a serious of behavioral oddities. Here are some, ah, common problems to, um, avoid.

- Speak loudly and clearly. Strive to avoid "ah" and "um" as filler between thoughts. Identify and eliminate any tendency to use certain phrases repetitively. Surprisingly large numbers of speakers have some phrase that they lean on too heavily in talks. "As you can see" is a perfectly fine phrase, but it gets annoying if you use it to preface every change of slide.
- Exercise pointer control. Use a pointer to point to specific objects on a slide when needed. Well-designed slides should need a minimum of attention from the pointer. Do not make sweeping arcs in virtual space or on the screen. If the pointer is a wooden or metal stick, avoid rubbing it on the screen. These behaviors are distracting.
- Many speakers discover that there is no good place to put your hands during a seminar. The best option is by your sides or in modest hand gestures to engage your listeners. Options to avoid include playing with change in your pocket, tapping on the desk, making wild or repetitive gestures, or fidgeting with clothing. If you know that you have a tendency to put your hands in your pockets while speaking, then imagine that you are going through airport security and *remove all metal objects.* The ringing of keys and coins will drive your audience to distraction.
- It is a good public speaking technique to move around the stage/floor in order to engage and speak to the various parts of the audience. However, do not pace the floor like a caged tiger. Too much movement begins to distract the audience. At a minimum, you should step away from the podium or screen a few times during the talk and move to address each part of the audience (left, center, right). Make eye contact with all parts of the audience. Even if you are relying heavily on notes, you can look up and make contact with all parts of the room. Above all, do not talk to the screen or use your slides as notes.

### The Practice Talk

The only way to bring all these pieces of the presentation together and to discover your own personal quirks is to give several practice seminars. We don't mean showing the slides to your roommate or reading the notes to a friend in lab. You must give the full seminar under game conditions. Assemble a willing group of faculty and students, and present your seminar just as you plan to do during the interview. Many departments have informal "brown bag" research groups that make an ideal audience for a practice seminar. Ask your colleagues to be frank in their evaluation. Hurt feelings are a small price to pay for constructive advice that improves your seminar.

Practice presentations are the only way to fine-tune the duration of the talk, identify problem spots, weed out behavioral quirks, and gain experience with possible questions.

## Plan for Emergencies

What will you do if the bulb blows or the computer freezes? If the mishap is likely to be corrected soon, try to keep talking as if the slides were still projecting. You can even quickly draw a graph or table on a blackboard (if there is one). This might be a good point to be sure everyone is still following the significance of your talk. You might ask if there are questions at this point while you are waiting for a new bulb. Whatever you do, do not just stand there looking like a deer caught in a car's headlights. If the problem is a more substantial one, you should probably inquire whether the search committee would like you to continue or wait until repairs can be carried out. If delay is not an option, you will have to make do without slides. Be sure you know the sequence of information in your talk even without slides to prompt you.

## At the Interview

### Check the Room

Chances are this will be the first time you will give a talk in the seminar or lecture room in this university. Not unlike a gladiator of old, you will find yourself in a pit surrounded by a throng of unfamiliar observers. You can help yourself out by learning about the room beforehand. Ask your host if you can see the room prior to your talk. This will help you make sure everything is in order before you enter the room later. Where are the lights, and can you control them? Will you control the slides? Is the room large enough that you will require a microphone? Will you need a glass of water, a reading light? Is there a podium or stage? Is there a clock in the room? (It's a good idea for you to have your own.) Make sure there are no obstructions or pitfalls to impede your movement as you interact with your audience.

### Thirty Minutes to Showtime

Odds are good that your host will have scheduled free time for you in the thirty minutes prior to your talk. This is not the time for you to start tweaking your talk or moving around slides (although we have seen this happen many times). Follow whatever routine you find effective for focusing and getting ready to perform. Ideally you will be feeling a little productive nervous energy. To keep this energy productive:

- Check your attire, wash up, and get your game face on.
- Review your notes to make sure you are organized and have everything you need.
- Take a walk and get some fresh air.
- Check the room one last time, and visualize how you will move and interact with the audience.
- Grab some coffee or a soft drink (as long as it won't make you burp).

### Beginning the Talk

As any sprinter or swimmer will attest, the start of the race is critical for success. It is imperative that you get off to a strong, clean start to your talk. It may help you to realize that many of the audience members are apprehensive at this point as well. Some of them have invested a large amount of time in the job search and are eager to see what you can do. The members of the search committee desperately want you to do well because you represent one of their choices. And the rest of the department wants to evaluate you as a future colleague. Your first couple of minutes will go a long way toward alleviating these concerns, so it's important that you have this part of your talk scripted (in your head or on paper) completely. Here are some suggestions to get off to a good start.

#### Begin with the Lights On

Let your audience know that you are comfortable and do not need to hide in the darkness. Make eye contact and engage your audience. Establish a presence before the room goes dark. (Oops! Did you forget to find out how to operate the lights before the seminar?)

#### You Had Me with "Thank You"

After your host introduces you, stand up and in a loud voice say something along the lines of "Thanks for that nice introduction, Dr. Greene. I would like to start off by thanking the department and search committee for inviting me to Major State University for this interview and seminar. I have enjoyed meeting the faculty, staff, and students and look forward to talking to more of you later." You should be directing your comments toward the audience. Try to make eye contact with folks in different parts of the room. At this point, you will want to tell your audience what you intend to do for the next forty-five minutes so they know what to expect. For example, you could start your seminar with the following: "I have three parts to my talk today. I will first provide a general introduction to my research program. Next, I will discuss the details of the major research project I am involved with now. Finally, at the end of the seminar, I will briefly discuss where I see my research going in the future and specifically what I would work on should I be offered the position here." This approach shows that you have taken control of the situation by telling the audience that you know what you will be doing and what they are in for.

*Here's the Beef*

At this point, you can begin a general introduction to your research program. Remember that not everyone in the audience has an intimate knowledge of your CV and supporting application materials, so take a few minutes to provide background for the coming talk.

**Closing the Talk**

Fight the temptation to end with a sunset photo. End with a strong series of statements that summarize the importance of your research and where you are headed in the future. As with the start of your introduction, these last three or four minutes should be memorized and presented cleanly. End your talk by saying something like "I would like to thank you for your attention," or simply "Thank you." Do not end with "Um, I guess that's all?" nor even with the harmless-sounding "Thanks and I will answer any questions." There are at least two reasons why this last phrase is not advisable. First, most scientific audiences are trained to applaud a speaker after they finish talking and again after the questions-and-answers session. By you bringing up the issue of questions, you make the audience feel unsure if they should clap or say something. Second, by convention, it is the role of the host to stand up and say, "We have time for questions if anyone has any." As we said earlier, find out who normally fields the questions in departmental seminars.

## Questions and Answers

How you handle questions is critical to the success of a research or teaching seminar. This is the one and only aspect of your seminar over which you do not have total control. Many departments have faculty members who pride themselves on asking obtuse questions just to confuse the speaker; these cobras are positively giddy in the moments leading up to a job seminar. Here are some things you can do to enhance your performance in answering questions. First, take a page from the politician's playbook and identify the major themes or take-home messages of your research and your strengths as a candidate. Try to work these into your answers when possible. Second, there is no substitute for practice. Practice your talk in front of live audiences as often as possible, and have them ask tough questions. Ask for feedback on your answers. Presidents do not go into debates without a series of mock debates beforehand. Do the same for your job seminar. Third, listen carefully to the question. It is surprisingly common for candidates to answer a question other than the one posed to them. You might want to bring pen and paper so you can keep track of those complicated multi-part questions, as well as write down any fantastic new ideas that crop up in the midst of the question period. Keep your answers on point, and resist the temptation to ramble. In the absence of a cue to stop, many nervous speakers keep talking long past any effective answer to the question at hand. Finally, we suggest that you make eye contact with the questioner. If the room is small, it is often effective to address your answer directly to the questioner. However, in a larger room, it is a good idea to repeat the question for the audience and direct your answer to the group as a whole.

No one knows your research better than you, so you should be able to anticipate some of the questions you will receive. If so, you might consider putting some extra slides at the end of your presentation to help answer these. As we mentioned earlier, you will look like a cool customer indeed if a question is asked and you can say, "That is an excellent question, and if I can have the slides back on, I will show you the answer." Finally, it is kosher to say that you don't know an answer to a question, but it's recommended that you do not do this for all questions.

# 9

## Social Time

> "*True story: After a pleasant dinner at a local restaurant, John was being taken to the home of a faculty member for an evening reception. He asked the driver if he would mind stopping at the liquor store. The driver said that it was not necessary to buy anything for the host, but the candidate insisted. He ran into the liquor store, bought a fifth of vodka, and offered some to the driver (a teetotaler), who declined. Once at the party, he held on to the bottle, finishing it off single-handedly over the course of the evening.*"

So why did this candidate not get a job offer?

A. He refused to share with others.
B. The vodka he bought was cheap swill.
C. His slurred speech at the reception made it difficult to understand what he was saying.

There is no doubt that interviews and seminars can induce a lot of stress. Fortunately, all this is tempered with some more enjoyable events—receptions, meals, and sometimes even outdoor activities, such as an easy hike or, for field biologists, a visit to nearby research sites. Social activities provide a chance to talk about science with students and colleagues in a less formal setting, to find out what life is like in the town where you may be spending the next several decades, and to relax and enjoy yourself. Your hosts also appreciate the chance to get to know the "real" you in a casual social environment. However, make no mistake about it, you are still being interviewed. During one of our own job interviews, the search chair said just prior to a social, "Relax, the interview is over now; we're just here for a good time." Nothing could be further from the truth (even if the chair really believed this to be true). During a job interview, you are *always* being evaluated. A night of chugging the local brew and hitting on attractive graduate students will get you noticed and talked about, but not hired. There are many rules (especially for meals) to ensure that you and your hosts feel comfortable and that you come across in a positive light.

In our experience, close hiring decisions can sometimes come down to particularly favorable or unfavorable impressions of candidates during social events. After spending the day talking about your teaching and research, social events give you a chance to show off other parts of yourself. Of course, most of us in academia are pretty excited about our work. If you are confident and excited about your work during the day, but over dinner you show no interest in talking about science, your hosts may wonder about your commitment to a life in academia. On the other hand, if you are incapable of talking about anything other than work, they may find you a bit narrow (not to mention dull). Do not miss the chance that socials provide for your hosts to get to know you from a slightly different perspective.

### Meals

At the risk of causing flashbacks to the family dinner table when you were ten, it is worth starting with the basics for mealtime during an interview. Remember, there is always the chance that some faculty member will be strongly influenced by what others would consider minor violations of dress conventions or manners. Meals, of course, offer many opportunities to show your worst (or best) manners. To some, these rules seem to make something that we do several times a day just too complicated. But there are good reasons for them. In *The Hungry Soul*, Leon Kass notes that good table manners "show consideration for the comfort and pleasure of one's fellow diners" (152) and promote community through shared rituals. In fact, Kass would go so far as to argue that table manners "make [the immediacy and intimacy of life] possible" (153).

You have just finished a very long day—giving a job seminar, interviewing with faculty members, meeting with students—and what you might most like to do is go back to your hotel room and watch *Gilligan's Island* reruns. But your host asks you where you would like to go for dinner (supper for our southern friends). The best thing to do is to suggest that your host decide. If you have any dietary restrictions, be sure to let those be known. If you are offered a choice (say, either Chez Beaucoup d'Argent or a fast-food joint), do not be afraid to choose the pricier of the two (and no, going back to your hotel alone and ordering room service is not an option). The department usually has a fund to cover these expenses, and faculty enjoy the opportunity to eat a good, free meal while getting to know you better. At smaller schools, be sensitive to the possibility that the faculty may have to cover your meal costs.

DON'T WORRY, PROF, THIS ONE'S ON ME.

> "In our department, faculty have to cover the costs of dinner when we take out visiting job candidates. It's not a deal breaker, but boy does it bug me when the candidate orders the most expensive dishes on the menu and the most expensive glass of wine!"

Once at the restaurant (and plan on showing up a few minutes early if you are making your own way there), a good rule of thumb is to follow your host's lead. Allow your host to sit first unless your host motions you to sit. If you are wearing a suit jacket and others are as well, leave the jacket on unless your host suggests otherwise. Once seated, do not look at the menu until others have done so. They may want to chat with you for a while before ordering. When asked what you would like to drink, do not order alcohol unless others do so as well. In any case, keep your alcohol intake down—you want to stay sharp. If you are not sure if it is okay to drink, you can ask your host if they usually have a drink with the meal during these occasions.

Be aware that there may be someone at dinner who does not look favorably on excess alcohol consumption. Even if another is encouraging you along ("Hey, let's have another round of tequila shots!"), practice moderation. In any case, do not get out in front of your hosts on the alcohol consumption (don't start dinner with a hearty "I'll buy the first round").

When ordering, you should only order an appetizer or dessert if others do, but do not feel compelled to do so. Avoid ordering the most expensive item on the menu (unless others do). After your host has ordered appetizer A, entrée B, and drink C, avoid saying, "I'd like what she ordered." And don't be afraid to try the local fare. If you are interviewing in Georgia, better to say, "I've never had grits, so I'll try them," rather than "Grits? Don't they feed that to hogs?" However, be prepared to eat whatever you are adventurous enough to order.

One of the challenging aspects of these dinners is to figure out how to eat your meal while being asked lots of questions. Order food that is easy to eat and not too messy (spaghetti with tomato sauce, unshelled crab, and a slab of barbecued ribs are all likely to get you into trouble). Make sure you leave enough space for talking between bites so that you don't choke on your food or end up displaying your partially masticated salmon to the entire table. Eat slowly, and put your knife and fork down from time to time—this is not a race to the finish. These are all fairly obvious things, but the specifics of table manners can be tricky. We suggest that you brush up on your basic table etiquette prior to your interview, especially if you are a recent graduate who has been surviving on pizza for several years.

Perhaps the most important determinant of a successful meal is good conversation. This takes focus and considerable mental energy, so be prepared and do not let a long day of interviewing catch up to you now and make you a dull dinner companion. Take an interest in your hosts, and divide your attention evenly among all your hosts. You do not want them to think that you only like to talk about yourself. But they are obviously interested in getting to know you better. It is appropriate to talk about non-scientific as well as scientific matters, but there are certain topics that you should consider off-limits. Do not talk about politically charged topics. Even if your hosts are revealing their red or blue leanings, it is best not to trot out your latest joke about conservatives or liberals. Also avoid talk about sexual matters (unless, of course, you work on mate choice, fertility, or a related topic!), and do not engage in off-color humor. If the meal is after your seminar, you might get some pointed scientific commentary or criticism. Maintain civility in your responses. In general, be a modest but friendly guest.

One word of warning: You may encounter atypical situations during some interviews. During one interview, one of us was dropped back at the hotel at 5 P.M. with no mention of meals or guidance for evening activities. On another occasion, the department head took one of us back to his house for a meal with his family (no students or other faculty attended). Be prepared for the unexpected.

## Receptions

Parties are supposed to be fun, so try to relax, be yourself, and if you are not too exhausted at this point, have fun. But remember, you are being interviewed and there are some important rules of behavior.

### Dress Reasonably

Just because you are at a social event is no excuse to violate basic rules of dress for a job interview. Do not be too casual or too sexy. As an academic, you have a bit more flexibility in your sartorial decisions, but don't get carried away.

### Easy on the Liquor

If alcohol is served, feel free to partake, but nurse your beer or drink through the evening. There is no quicker way to go down in flames than to drink too much during an interview. On the other hand, if alcohol is not served, do not request it.

### Be Sociable

The reception is often your best opportunity to talk with people who were not part of your scheduled interviews. Do your best to meet with as many people as possible, even if only briefly. Do not limit your time to only the people that you think are important—the chair, the search committee, the big-name scientists. If you find yourself cornered by the departmental bore for twenty minutes, do what you can to extricate yourself in the most polite way possible: "I think I'll get another drink" or "Can you tell me where the washroom is?" are surefire ways to shift your social circle. Having your time monopolized by one or a few people is a common pitfall at socials, and your hosts may not rescue you (they will be happy to park by the free food and chat with friends). Take responsibility for your own circulation at the social event.

### But Don't Be Too Sociable

Do not flirt with students. Or with the department head's spouse. Or with the department head. Check yourself to make sure you are not unconsciously hanging out all evening around the most attractive people in the crowd.

### Talk to Students

If you are interviewing at a department with a graduate program, make sure you talk to graduate students as well as faculty. There may also be undergraduates at the social event, and they should get some of your time as well. If necessary, go out of your way to engage students in conversation. They may form their own quorum off to the side (strategically located between food and beer). Join it for a while.

### Scientists Talk Science!

Do not be afraid to engage in scientific discussions, but be careful that you don't insult your hosts. You may have a very strong disagreement over a scientific issue with someone, but be civil. And avoid getting in any arguments over politics or religion. Even if you think you share opinions with the majority of the people at the reception, you risk insulting someone who may have a critical vote when your name comes up at the next faculty meeting. We have also seen that as members of the faculty and students finish off their second or third drink for the night, sensitive departmental matters may come to the surface ("Guess what that idiot chair did during my tenure decision!"). Pay close attention to these sorts of conversations, but be diplomatic and do not pry into these issues.

### Don't Forget about Tomorrow

Do not stay up past your bedtime. You may have additional meetings in the morning. It is perfectly appropriate to let your host know that you need to get back to the hotel at a reasonable hour, and once that hour is reached, to ask if someone could take you back.

### Outside Activities

Sometimes you will be asked to do something a little out of the ordinary. For example, graduate students may take you on a hike or to see local field sites (as long as you don't consider your gene sequencer the field), a concert, or a sporting event. As with all other events, you are still being observed. Be gracious and enthusiastic about the event, and show interest in your hosts. Dress can obviously be more casual in these situations.

**10**

# The Negotiations

*You can't always get what you want*
*But if you try sometimes you just might find*
*You get what you need*

MICK JAGGER & KEITH RICHARDS

If you have followed the advice in this book, chances are the phone call will eventually come. You have been offered a job! The next few hours will be a euphoric blur. There are congratulatory phone calls and e-mails from friends, colleagues, family, and even from members of the department trying to hire you. You can tell your parents that those two decades of college and postdocs finally paid off. You can perceive awe in the glances of graduate students in the hallways. At long last you have punched your ticket to the ivory tower. Slowly, this euphoria will wear off, and you will be faced with a new reality: negotiations. This is a tense moment. Most of us have little or no experience with this sort of thing, especially in the case of our first job.

Furthermore, if you have targeted your job search to situations that are well suited for your career goals, you want this job! You will be tempted to settle for less than you can legitimately negotiate.

The good news is that your job negotiations will usually be a lot easier than negotiating in the business world. In academia, you are unlikely to encounter the adversarial approaches that sometimes occur in the corporate world. Furthermore, in most cases the department head with whom you will be negotiating is on your side. It is in the chair's interest to see that you get the resources you need to succeed in the first few years of your career as an independent scholar; he or she wants a happy, productive new faculty member. Thus, treat your negotiations with the chair as something of a collaboration to maximize your success and strengthen the department. Do not be too complacent about your negotiations, though. You need to think carefully about what you need to succeed, what you need to be happy, and what you would like in the best of all possible worlds. Many of the things you negotiate at this time cannot be readily changed in the future (e.g., lab space, salary, equipment). Getting it right at this stage can make a big difference toward your long-term success.

We offer a few suggestions to help guide successful and productive negotiations, but our first piece of advice concerns what not to do. Although you are just one tiny step from accepting the job that you have spent many years training for, avoid the temptation of just muttering a grateful "Yes, thank you, thank you" when the dean or department head tells you what they plan to give you. It is much better—and your potential employers will certainly expect—for you to say that you will give the offer serious consideration. Although a quick "Yes, I'll take it" would save you the time of reading this chapter, let's assume that you are willing and able to negotiate. Here are some guidelines on *how* you should negotiate and *what* to negotiate.

### How to Approach Negotiations

Be reasonable. First and foremost, you must realize that what is negotiable in terms of salary, space, and start-up will vary widely across the academic landscape. Remember that what seems fair and reasonable to request at a large research institution may elicit shocked silence from the department head or dean at a small teaching school—or even at a cash-strapped regional university. Fortunately, if you have taken the time during your graduate training to understand how your career aspirations fit into the types of jobs available, you will already know what you can expect at different types of schools. To move beyond this basic understanding, explore what job packages typically look like at comparable schools in the same region. Bear in mind that within the life sciences, different types of jobs in biology will be offered very different packages. A molecular geneticist, a field ecologist, and a bioinformatician will obviously have very different start-up needs and can command quite different salaries (for example, at some schools a bioinformatician may command a salary that is 10 to 25 percent greater than that of other biologists). Large negotiable start-up packages are common at large research institutions, whereas smaller, fixed start-ups can be encountered at smaller institutions. There are even institutions that do not provide formal start-up packages (although they will try to include your most important needs in the year's budget).

Be realistic. Once you have done your background research, set realistic goals. What are your minimum requirements? If you do not have the resources to do what you want in your career, are you prepared to walk away from the table, perhaps even putting off accepting a job until next year? Consider possible trade-offs between different elements in your overall package. For example, are you willing to take a salary below what you want (but not below what you *need*), in exchange for twice the start-up funds you expected to receive in your best-case scenario? If you do not know where the balance point is between what you want and what you minimally need, then you are entering negotiations at a real disadvantage.

Be patient. Sometimes the process of negotiating an offer can take just a few days. More often, it takes weeks or even months. If your needs are especially complex (a position for your spouse, a new greenhouse), the department head may need to negotiate with higher-ups in the administration, and this can take some time. Do not formally accept the offer until all major issues have been negotiated to your satisfaction. However, the negotiation process is not completely open-ended. You will sometimes be asked to respond within a particular time frame. Do not be forced into a hasty decision, but understand that departments can have a legitimate interest in expediting negotiations. As the job season wears on, the pool of viable applicants gets smaller. If you keep a department negotiating for months only to decline, the department may have to settle for an inferior candidate (or even lose the job line altogether).

Be fair. While the process may take a good while, it is critical that you negotiate in good faith. This can mean several things. What do you do if you have multiple offers? If school A offers you a better salary than school B, but school B provides better start-up funds, there is nothing wrong with asking school B to match school A's salary offer (but be prepared to provide a copy of the written offer from the other schools). Of course, if you have no intention of taking the job at school A, do not drag their negotiations on, knowing that you are going to accept an offer from school B. If you tell a school that the only thing stopping you from choosing them over another school is their salary, and they then match your request, do not come back with yet another element in the overall package that you want changed. The department head will only go to bat for you so many times until he or she comes to think that you are not negotiating in good faith. If you come back to the table for seconds and thirds, your prospective department will become frustrated and may begin to think that you are going to be a high-maintenance colleague. The first impression you want to give your new chair is that of someone who knows what he or she is worth, but who will also be a fair and cooperative department citizen.

Finally, be honest. If, during your interview, you tell your colleagues that you would be more than happy to teach an introductory statistics course, do not look shocked when you find that they have assigned that exact course to you. If you cannot teach a course in biostatistics, don't claim that you can just to impress the department during your first visit.

## What to Negotiate

No discussion of what to negotiate is complete without reiterating that across the academic landscape, there is enormous variation in what universities are willing and/or able to offer you. Before you begin to negotiate, you need to do a reality check (and a little homework). Depending on the nature of the university, the region, and the type of position, salaries for starting faculty can vary by a factor of two and start-up funds can vary by two orders of magnitude. One size does not fit all. While you are seeking advice on which ballpark you need to be in, remember that your advisor or supervisor is familiar with only a narrow slice of the academic landscape; he or she will not always provide you with the most realistic or helpful advice. In fact, let's be honest: your major advisor can sometimes provide horrible advice. We have repeatedly seen advisors at major research institutions offer advice that did not serve an applicant's best interests during negotiations for jobs at midsize and smaller schools. Get advice from people who know the environment in which you are seeking a job.

So let's get down to brass tacks. After giving an offer "serious consideration" (high fives and champagne in the lab), what now? There is a lot to think about, but we can simplify the process. We break negotiations down into five distinct elements—salary, space, start-up, time, and job duties. As an acknowledgment of the variation mentioned above, we will try to give you a feel for the high and low end of the range of possible negotiations.

### Salary

It feels especially hard to negotiate for a higher salary. We associate our pay, and how it compares to others, with our self-worth. Your friend from college who studied finance and did not even know how to wash his own clothes may already be earning more than you can ever hope to earn. Do not worry about it—what we give up in salary as academics, we get back in intellectual stimulation and academic freedom (if you doubt this, go back to chapter 1 and reevaluate whether you really want to be a biologist, much less an academic). However, there is nothing wrong with earning a good salary, and negotiations are the time to set the bar as high as possible. Recall that increases in your salary once you are hired will tend to be smaller than the year-to-year jump in salary for new hires. This is a phenomenon that leads to problems of "salary compression" or even "salary inversion" for current faculty. Your colleagues may become disgruntled about the situation, but they are likely to blame the department head, not you. So you will want to establish the best starting point you can.

Unfortunately, of all the negotiable elements for a new job, salary may be the least flexible in an academic setting. In some systems (like the University of California system) the salary is set according to a relatively objective scale, and as a newly minted assistant professor, you will have little leverage to deviate from that scale. Salaries will vary across different kinds of institutions, with salaries typically being greater at large research universities than smaller teaching colleges. Private institutions usually have salaries similar to public ones, though within the public system, average salaries are higher at universities where faculty are represented by unions (2003–2004 CUPA-HR study, as reported in the *Chronicle of Higher Education*, May 7, 2004). Table 10.1 shows some general numbers, averaged across disciplines, to give you an idea of what typical academic salaries look like at Ph.D.-granting (tier I) schools. They also include disciplines that typically earn less than faculty in the natural sciences.

In the cases where there is some flexibility, you can do a few things to maximize your chances for a successful negotiation. As we have already

**Table 10.1. Average Salaries in Doctoral Institutions, 2004–2005**

| Rank | Public | Private |
|------|--------|---------|
| Assistant | $58,310 | $70,640 |
| Associate | $68,576 | $82,456 |
| Professor | $97,948 | $127,214 |

Numbers from a survey conducted by the American Association of University Professors and published at http://www.insidehighered.com/.

mentioned, start by having an idea of what comparative starting salaries are for jobs in the same discipline and at similar types of schools in the region. It does not hurt to have these numbers in mind even before your first visit, as you may be told salary ranges at this point. Better to be prepared for what is coming, so you do not respond either by showing great disappointment or screaming like a Lotto winner. Here are a few general guidelines:

- Do your homework. How much do you want, how much do you need, and how little will you settle for? The answer to these questions may differ depending on the location of your job offer. According to one on-line cost-of-living calculator (http://www.datamasters.com/), an offer of $60,000 in Tucson would give you the same standard of living as an offer of $140,000 in New York City. Compare this number to the salaries offered by similar schools in the region. There are several websites that provide average salaries (such as the sites of the National Education Association, http://www.nea.org/he; and that of the *Chronicle of Higher Education*, http://chronicle.com/). For public institutions, faculty salaries are in the public domain.
- Do not be the first to put a number on the table. When you hear the first number, do not say "OK." Before telling you what the school is prepared to offer, the chair or dean may ask what your salary requirements are. Back away from this question politely but firmly. You might counter with something like: "Why don't you give me an idea of the range of salaries that your school pays for this type of position. We could then work out where in that range I should be, given my credentials." Mind you, if the bottom of your range is higher than the top of their range, you can counter with that. Alternatively, if you are not yet ready to negotiate (perhaps you still do not have an offer in hand), you may tell them that "I am much more interested in the research opportunities here than

I am in the size of the initial offer." If they persist, you might take the advice of Noel Smith-Wenkle (http://www.nmt.edu/~shipman/org/noel .html). On receiving a second request to name your salary, you can say, "I will consider any *reasonable* offer," and if asked yet again, you might say: "You are in a much better position to know how much I am worth to you than I am."

- Be positive. Remember that the department head is on your side. If he or she is clearly *not* on your side, you might want to consider what this means about becoming a member of this department. That said, it is not worth going to the mat with the department head over a few hundred dollars, but a few thousand dollars can make a significant difference.
- Find out whether the offer is for a nine-month or twelve-month salary. For nine-month salaries, find out if the school will be willing to pay part or all of your first year or two of summer salary, until grant funds start coming in. Most research institutions will offer you a nine-month salary, under the assumption that your summers will be devoted to grant-funded research, rather than teaching. If you are offered a nine-month salary, you will be able to pay yourself during the summer from grant funds, or you may have the opportunity to teach summer courses in exchange for supplementary salary. Rules about summer salary from grants may vary. While NIH will pay for a full three months of summer salary, NSF will only pay for two of the three summer months. In any case, if your offer is for a nine-month salary, don't forget to add between 22 and 33 percent to calculate your annual gross income if you are able to bring in summer salary. Of course, the advantage of twelve-month salaries is that you don't have to worry about the possibility that you might have a year or two with no grant funds coming in. In a more teaching-oriented environment, be sure that you inquire whether summer teaching is available to cover summer salary.
- If you are negotiating with a medical school or a private research institute, they may expect you to pay some or all of your annual salary from grant funds within a few years of starting your new job. This is known as a "soft-money position." Make sure you know the full terms of what the school expects from you, and how much of your salary they can cover if your grants won't pay 100 percent of your salary.
- Know when to walk away. You must have in mind some minimum salary below which you are simply not prepared to accept. If a school cannot match this basic level, be prepared to walk away, even if this means

putting off your job search until the following year. If you looked good enough to land an interview this year, you should be even more attractive next year (as long as you heed our advice and publish).

- Know when to run. If a school cannot meet your minimum demands and you know that you will go elsewhere, do yourself and them a favor by turning down the offer quickly, giving them the opportunity to offer the position to another candidate.

## Space

Determine what your space needs are ahead of time. On your first interview, it is appropriate to ask to be shown office and lab space. Now is also a good time to ask about other research space, such as greenhouses, animal care facilities, field stations, and so on. Make sure when you negotiate for space that the space will be available by the time you arrive. One person we know showed up at her new job to find that the promised office and lab space was not available, and the only available space was in a completely different building, well away from her colleagues. Be wary of claims such as "We do not yet have free space, but we will have something available by the fall." If space is promised you, get it in writing. However, be aware that space availability often depends on moving existing faculty, budget allocations, and action by college or university-level administrators. Thus, be sure that you and the department chair reach a clear understanding about what can be guaranteed and what can't. Be sure that your space is connected to the local computer network.

The negotiation stage is the time to ask for any remodeling that you may envision. At this stage, it is relatively easy for the powers that be to move walls, doors, benches, or cabinets, if not mountains. Once you arrive on campus, you are unlikely to receive the same attention. Be sure to determine how renovation will be paid for: By the university? By your department? Out of your start-up funds?

### The Details

In negotiating for space, there are several important criteria to keep in mind.

- Where is your office in relation to your lab? One of our colleagues has an office in one building and a lab in another one. This can be good for avoiding bothersome graduate students, but some of us find that graduate students make fun and stimulating colleagues.

- Is your office close to colleagues? You may want to be located close to a potential collaborator, but not too close to the resident gossip.
- Lab space often consists of more than just the lab. What about storage space for boxes and currently unused equipment and supplies? Space for incubators? A media prep room?
- Is there office space outside of the lab for graduate students? While you may not be bringing grad students with you to your new job, you will want to make sure that when they start applying to work with you, you will be able to show them space that will make them want to join your lab.

### Start-up Funds

Start-up offers can vary enormously from one school to the next. A smaller school may offer you a few thousand dollars and some older equipment left by recently retired faculty. A tier I research university may provide over a half million dollars' worth of support. What should you do?

- Do your research ahead of time. As always, you need to come prepared. Talk to other postdocs and faculty at your current institution. Put together a "minimum required" list of what you would need to do your research. These should be the "non-negotiables." Then put together your ultimate wish list. This should include everything you would like to have in your lab, without being frivolous (the big-screen TV and portable beer cooler are not likely to go down well with your new procurement officer). For both lists, spare no detail. Larger items—thermal cyclers, computers, microscopes, field vehicles, et cetera—are obviously important to include as you add up how much money you will need. But don't forget the smaller items that can quickly add up—Kimwipes, centrifuge tubes, data loggers, gloves, a tub of agar, a coffeemaker. You may take these items for granted as a grad student or postdoc, but someone is paying for them.
- How flexible is the start-up package? There are two ways in which start-up funds can be more or less flexible: What can you buy, and when can you buy it? Some schools may give you a generous amount of money for start-up but require you to spend the bulk of it within your first year. You may want to negotiate a longer time frame. This is especially im-

portant for computers, which go out-of-date quickly. (For computers in particular, wait until the last possible moment to buy what you need— you will get more bang for your buck.) You may also be told that a fixed amount of start-up is available for supplies and expenses or staff, and the rest must be spent on equipment.

You may be able to work around both of these constraints. Of course, just how constraining these limitations are will depend on your own needs. You may find that your start-up is generous enough that you can buy the largest pieces of equipment you need and still have plenty left over for supplies, expenses, and staff during your first couple of years. But if this is not possible, do not despair. There are other ways to facilitate these large purchases. If you are fortunate to already have a substantial research grant in hand, this can be a bargaining chip in your favor. Having already proven that you can bring in grants, the university can be reassured that a dollar invested in you will pay dividends in the future in terms of the overhead that you bring in.

What will you get? The actual start-up package can be used for a variety of different things, and these can be negotiated individually:

### Equipment

Equipment (usually items over somewhere between $1,000 and $5,000, depending on the school) will need to be purchased through a procurement office (computers often excepted). For some readers, you may already have experience with this in your current lab. But don't be surprised when you are inundated with salespeople keen to convince you why their company makes the best microscope, or thermal cycler, or imaging system, or whatever it is that you are looking for.

In some cases, the equipment you need may already be on campus. You only need to convince the owners of the equipment to share (in this case, your prospective department head may serve as a good ally, working out an agreement that satisfies all concerned). You may also inherit a perfectly good piece of equipment from someone who has recently been lured away to another institution. Make sure you have proper service support for "pre-owned" equipment.

If you are interviewing at a smaller school with smaller budgets, be careful when negotiating for equipment. Your request for a $70,000 sequencer may convince your prospective employer that you do not have a clue about the nature of the school at which you are interviewing, and any chance of an offer will fly out the door. But don't give up on your dream machines. Instead, you might try suggesting something like "If I were to be offered a job here, I was thinking it might be a good idea to write a multi-faculty proposal to obtain funding for a gene sequencer. Given the interests of the faculty, I think we'd have a great chance at succeeding."

Don't worry yet about how you will actually spend your start-up. You might imagine that the day you arrive at your new lab, you will be greeted by your colleague down the hall offering to buy you lunch, graduate students looking to add your expertise to their dissertation committee, or keen undergraduates interested in new research opportunities. More likely, there will be a line of salespeople, looking rather out-of-place with their suits and briefcases. They have spotted a new faculty member, and that means money. Fortunately, they have a lot of expertise and will make your large equipment purchases a relatively painless experience.

### Staff

As part of your start-up package, you may be able to negotiate funding for technicians, students, and/or postdocs. Some departments will have excellent funding opportunities for students, in which case you will not need to negotiate on their behalf. In other schools, you may want to ask for anywhere between two and five years of guaranteed funding for one or two graduate students.

A technician can be especially nice to have around when you are looking for someone to help unpack boxes of new supplies, to set up equipment, and to train new undergraduates. This is especially the case if you are spending your first semester writing grant proposals or planning for courses that you have never taught before.

### Travel

You may not be able to put this on your lab bench, but don't forget this important element of your start-up package. Your success as an independent scientist will depend in part on your ability to interact with your colleagues. In the first year or two, as you are waiting for grant funding to come in, being able to travel to one or two conferences a year can be critical. Your department or school may have an in-house grant system to fund travel. If not, try to put some in your start-up package.

### Don't Forget the Small Stuff

It is relatively easy to calculate how much you will need to spend on equipment. When it comes to supplies, however, you will inevitably forget things. Find out if you can use start-up for office supplies or if the department supplies these through overhead return (the "tax" that the university takes on all federally funded grants). On your first visit, find out how faculty pay for items like dry-erase pens, printer cartridges and paper, paper clips, and so forth. How do they pay for local and long-distance phone calls? What about photocopying?

### Time

No matter how good a negotiator you are, no matter how fast you run, time will continue to move forward. And of course, the older you are, the faster it moves. But this is one time when you *can* slow the clock down just a bit. You may be in the middle of an exciting and productive postdoc, learning new techniques and cranking out manuscripts, enjoying the freedom from teaching and administration. If that is the case, you may be able to negotiate a six- to twelve-month delay in the time when you start your new job. However, in some cases this will simply not be negotiable. If a chair will not move the start date, it may be that he or she is constrained by the university administration or by the firm commitment to teach certain courses in the fall.

If you are coming to your new job from a previous tenure-track position, you may also want to negotiate how quickly the tenure clock spins. Normally, a new professor will go up for tenure after four to six years. But if you have already been an assistant professor for two or three years, you may want to get some credit for time served, and so go up for tenure early. For all faculty, make sure that you are clear on just what the expectations are for tenure, in terms of publications, teaching, administration, and grants. What you need to do to attain tenure is not a negotiable item, but it is nevertheless an important issue to be clear on from the start. For those with previous faculty experience, as with the tenure clock, be sure to discuss your "sabbatical clock," if your new institution has an official sabbatical policy. Will your new employer allow you to count your previous employment as years toward your sabbatical?

In the course of your negotiation, you may also want to ask that a senior faculty member be provided as a mentor to help make sure that you are on track as you move toward tenure.

### Teaching and Service

Once you have an academic job, you will suddenly be asked to add "teacher" and "administrator" to your career as researcher. And you will find yourself somehow managing to do all three of these things simultaneously. But as with most components of your offer, when and how much you do of each of these is sometimes negotiable. Academic jobs are typically defined by the proportion of research and teaching you are expected to do. For example, you may have a 60 percent research/40 percent teaching appointment. Now, this obviously adds up to 100 percent, but there is no mention of service here. This does not mean that you will be spared those duties. Let's just say that we know very few academics who only work forty hours a week.

You may be able to negotiate these percentages, but in most cases they are fixed. What you can expect to negotiate is what you have to do in the first year or two. For research-focused schools, they may give you a cut in teaching and administrative duties in your first year. This is done to give you more time to write and submit grants, hire technicians, set up your lab, and attract students and postdocs. You may be able to negotiate the specific cuts in teaching and administration that you get, but as with most other negotiated agreements, be sure to get it in writing. One of us negotiated a "zero service" load in the first year and so was able to graciously bow out of a request to sit on a committee in his first year, when he was asked by the new department head.

For the longer term, you may wish to negotiate particular courses you teach. In most departments, teaching loads are distributed evenly, so it is unlikely that you will be able to negotiate the number of courses that you teach. Bear in mind that any cut in teaching that you are able to negotiate might need to be covered by your colleagues. They are doing you a favor— be sure to step up to the plate when your turn comes around.

## Anything Else?

We have discussed five major issues that are up for negotiation—salary, space, start-up, time, and job duties. There is a sixth category of items that may or may not be negotiable and may or may not pertain to your own situation. Some of these are relatively minor items, like moving costs. Others, like child care or mortgage assistance, can sweeten the pot substantially. And one of the more complex and critical issues for some is that of the spousal hire. We will address this last issue in our next chapter. Finally, if you do have a partner, you may want to negotiate the time and money to bring them in to visit the city before sealing the deal. Some universities will comply; some have neither the time nor the money to be able to accommodate this request. If you are unable to accept an offer until your partner has visited the city, there is no guarantee that a university will cover these costs.

## How Do You Conclude?

Once you have negotiated each of these issues to your satisfaction and have a formal offer letter in hand, you can reply to the department head with a written acceptance. In that acceptance letter, it is appropriate to say that you accept the offer, with the understanding that the following elements are included in the offer. And then you can list the basics of your negotiated package, including the dollar amounts of salary and start-up, the teaching requirements, and so forth. You can then say that if these terms are acceptable to the department, they can consider this a formal acceptance. You don't need to include every gory detail, but if you really want to make sure that you have something in writing, now is your chance.

# 11

## *All in the Family*

Someday our world will have gender pay equity and universal day care, and presidents of well-known private universities will not claim that differences in mental acuity account for the fact that there are more men than women at the highest levels of science. When that day arrives, we will revise this book and remove this chapter. In the meantime, we hope that some of the issues raised here are of use.

Throughout this book, we have explored various strategies from the moment you enter grad school right through the job interview to ensure that you find the academic job that is right for you. By now we hope you are feeling a little more confident in your ability to land that job and not too daunted by a process that turns out to be more than just shipping a CV in the mail. Until this point, we have shown you that while this entire process may not be easy, it is pretty straightforward. But even in our enlightened times, one set of complicating issues is still likely to be a cause for concern at some point along the way—gender and family.

Consider some thought-provoking data from a study carried out by Mary Ann Mason, the dean of Graduate Studies at UC Berkeley (the full text of which is available online at http://www.grad.berkeley.edu/deans/ mason). Mason analyzed data from 160,000 people who received their Ph.D. between 1978 and 1984. Bearing in mind that this group falls at least a generation before the expected readership of this book, the results are nevertheless rather sobering. Mason found that women in this cohort who had children were 30 percent less likely to land a tenure-track job than those without kids. Among tenured faculty, women were more than twice as likely as men to be single and without children (26% vs. 11%). And among faculty who had children early in their career (defined as within five years of earning a Ph.D.), 53 percent of women achieved tenure versus 77 percent of men. In contrast, for women with no children or children late in their career, 65 percent achieved tenure. Among tenured faculty, while 70 percent of men were married with kids, only 44 percent of women were.

While there are still plenty of problems with our institutions when it comes to women in science, we believe that things are far better now than they were for those receiving Ph.D.'s twenty-five years ago. Many universities now have programs for mentoring women at all stages of their career, flexible tenure clocks to accommodate childbirth, in-house child-care programs, and more. Come to think of it, these are all good things to ask about if you meet with the dean or during your negotiation process should you receive an offer. Even if your potential employer does not yet have these policies implemented, they will receive a healthy reminder from you that these issues matter! In any case, we believe that you *can* balance a successful scientific career with a rewarding family life.

In the rest of this chapter, we explore some gender- and family-related issues that may arise, from graduate school right through the negotiation process. Each of your experiences will obviously be unique, but we hope that at least you will be aware of the issues that may arise. We will examine a series of issues, including (1) having a family, (2) finding the right mentors, (3) unfair treatment during the interview, (4) interviewing while pregnant or with small children, (5) the "two-body" problem, (6) women and negotiation, and (7) domestic partner issues.

## Family

Graduate students have asked us when the best time to begin having a family is for an academic. Based on several thousand years' worth of evidence, we believe that the best time to have a child is about nine months after conception. But seriously, there is no "right" or "wrong" time to have a child, other than when you feel ready to make the commitment.

Children are guaranteed to provide both desired and not-so-desired interruptions from work. There are clear benefits to having kids early on. You will have more energy, you will have a more flexible schedule, and by the time you are worrying about tenure, your kids will be in school and require less of your time. But there are also benefits to waiting to have kids until you have gone through the tenure process. You won't be worrying about the time required for child care interfering with your push for tenure, and you will probably have more financial resources. On the flip side, given that by the time you come up for tenure you may be close to forty, you also need to consider the biological challenges of starting a family later in life.

Ultimately, perhaps the best answer we can give to the question of when to have kids is simply "when it feels right." It is an intensely personal decision. Much as professors like to think they know what is best for their graduate students and postdocs, this is one decision that should be made by you and your partner.

## Mentors

Among the many women we spoke with who had successfully navigated the challenges of an academic career, we often heard comments like that from one senior faculty: "I was extremely fortunate in having Ph.D. and postdoc advisors . . . who treated their protégés fairly and evenly without regard to gender." When choosing an advisor, look for evidence that he or she has a record of training women with successful careers. In addition to your major professor or postdoc advisor, you may wish to seek out women as mentors who demonstrate an ability to balance a successful career with a fulfilling personal life.

It is important to find a mentor who will not only respect you as a scientist, but who will also support your life trajectory, even if it includes time for family or other activities. In this regard, look out for double standards. One of our female colleagues was told that she would fail as a scientist if she were not willing to give up the time that she spent dancing. But as she noted, "[My] lab mate who played basketball daily was never told any such thing."

A good mentor will help you to create a foundation of scientific experience upon which to build your career. Does your mentor help introduce you to other colleagues at meetings? Does he or she nominate you for awards? Are you encouraged to apply for grants?

While all these things are signs of a good, supportive mentor who respects you as a scientist, a good mentor will also make sure that you help yourself. As numerous women we have spoken to note, sometimes women are less likely to put themselves forward:

*"It is a recurring pattern I have observed firsthand over and over, that women start graduate school saying that they don't want a faculty job 'because they don't want that high-stress life style.' My suspicion, and it's been borne out several times at least, is it is not so much a straight lifestyle choice as it is a lack of confidence and low self-esteem. That is, they say they don't want that kind of job because they don't think they can do it . . . when really they are just as competent as the male graduate students I've seen. By the time they finish their Ph.D.'s, many of these women seem to see that it's actually not that hard to be a scientist . . . but I wonder how many great people talk themselves out of it early on."*

Many women are less likely than men to put themselves forward for prizes and awards. But as this whole book makes clear, a successful job search is predicated on the ability to unashamedly advertise your strengths. A good mentor will help you to develop this confidence.

One way to impress search committees is by the prizes that you have won. But one senior woman scientist we know who has been on numerous prize committees regretted the relative paucity of women candidates.

When choosing a mentor, bear in mind that you can have more than one. If your academic supervisors are men, or women who have chosen not to have a family, you may wish to actively seek out other "personal life" mentors at your institution.

## The Interview

You can be pretty sure that your research and teaching statement and your list of publications, awards, and grants will be evaluated without regard to

> "When women consider nominating themselves for an award, they will often compare themselves to past award winners. Indeed, they will often compare themselves to the most famous of the past award winners. Even if they do incredibly good science, this comparison will often lead them to think, 'I'm not THAT good.' This sort of checking one's contributions against the perceived value of others' contributions is, I think, partly to blame for why women are less likely to apply for awards. One flaw in this comparison is that a person might not match up to the BEST of previous award winners but still be in the running for the award in any one year. A second flaw in this comparison is that past award winners will have been doing science for longer, making a fair comparison difficult."

your gender or family situation. However, once you show up for an interview, you may indeed feel that you are not being treated fairly. It won't necessarily be more or less egregious than in any other setting in our culture, but, nonetheless, it pays to be prepared. Here we note some inappropriate questions or comments that you might hear and ways to deal with these situations. We will also discuss what to do if you are interviewing while noticeably pregnant or with a baby in tow.

In our informal and unscientific survey of women in science, we asked if women thought that their interviews had gone differently than they might have for male applicants. The answers varied greatly. One woman (who feels that she *has* been treated unfairly as a tenure-track faculty member) felt no discrimination during the interview process. Others spoke of being condescended to by older male faculty and, perhaps more nefariously, of the use of gender-specific language that could be perceived as a put-down. One tenured scientist was told that she "charmed the pants off the search committee." We think it unlikely that a male applicant would have heard the same comment. During a faculty meeting in one of our departments, a candidate was described as "vivacious." The comment was intended as a compliment, but such gender-specific language runs the risk of turning the focus away from the scientific merits of the candidate.

### Inappropriate Questions

"So, what does your husband do?"

In most settings, this may be a most innocuous question, a part of daily small talk. You may hear the same question during your interview, perhaps

asked by someone over dinner, and intended only as a friendly conversation opener. However, the question is not only inappropriate, but also illegal in the interview arena. Your marital status is irrelevant to your qualifications. How to handle such questions? There are several options:

- Emily Toth, the author of *Ms. Mentor's Impeccable Advice for Women in Academia*, suggests that you say something like "I do have a partner, a freelancer who'll be coming with me" or "I live alone," concluding that "the best you can do is to be dignified, precise, and good-humored." To which we would add: while the question should not have been asked in the first place, it's best to be honest. If your partner is an academic and you had chosen to hold off on revealing this information until later in the process, you may as well bite the bullet and hope for the best.
- Politely respond with something like "I'm not sure I understand why you're asking that particular question." As the interviewer responds, you should be able to determine whether he or she was just making small talk or had another motive. If you respond more directly with something like "Did you know that questions about marital status are actually against the law?" you might find yourself sitting through several minutes of awkward silence while you pick at your dinner salad.
- Try something slightly less threatening, such as "I'm not really comfortable talking about my personal life." Even here, you run the risk of putting off someone who is wondering about having you as a colleague for the next twenty-five years.

As one of our colleagues noted, "Interviews are geared toward people who are very confident at presenting and answering questions." Some old, sexist fossil in the department may upset you with a stupid question. The best way to deflect these questions is with humor. Try to hold off on venting your anger until you are safely back home. Once you hear from the department (with an offer or otherwise), you might then send a gentle note to the department head about this colleague, which should ensure that future job candidates do not have to endure the same treatment as you did.

### Oh, Mama!

What should you do if you are invited to interview at a time that you are nursing a baby or are visibly pregnant? First of all, remember that while an interview is your opportunity to shine, the same is true for the department that has extended to invitation. If the department can accommodate your needs, it will enhance its own reputation. One of our colleagues was asked to interview for a job during a time when her partner was away, so she had full child-care duties. When the invitation came to interview, she requested that the department provide a full-time nanny to care for her daughter (who was young enough that her travel costs were free). The cost to the department of hiring a nanny for two days was a small fraction of the other expenses. The department obliged, and while our colleague did not accept the job offer that she eventually received, she was mightily impressed by the department. However, this service may not be an option at a smaller school with limited funds.

So don't be afraid to ask for what you need. If you are nursing, let the person in charge of your schedule know ahead of time that you will need breaks to feed your baby or to express milk if your baby has stayed home.

### The Two-Body Problem

There was a time—we have seen the sepia-toned photos—when a job candidate was almost certainly male and the single breadwinner in a family. At the very least, one could be almost certain that an applicant did not have a spouse working in the same field. Now the academic landscape is full of dual-career couples. Although sharing an academic career with your partner is rewarding, you will face a number of challenges on the academic job market. The good news is that there are several ways to accommodate a dual-career situation, and we know of many happy, successful couples in academia. However, a happy ending will require (1) that you and your spouse agree on the career options you are willing to pursue, and (2) that you determine how and when you will raise the dual-career issue with a potential employer. These are both tricky steps, but we can provide some information that should help.

There are three options for the dual-academic family: (1) You each obtain independent positions at the same institution. This may occur simultaneously, or less ideally, you may obtain a position and then hope that a second position opens up for your partner sometime in the future; (2) you share a single position; or (3) you land a job at university A, your partner ends up at university B, and you both join a good frequent-flier program. We explore each of these in turn, but be aware that a good strategy for one school may be less ideal somewhere else. It is important to understand that there is tremendous variation among universities in how they will respond to dual-career couples (remember our chapter on the different academic job settings). Before you decide on a particular strategy at any given school, try to learn as much as you can about the culture of that university. Is there a good balance of men and women on the faculty? Are there signs of academic couples on campus? Any examples of shared jobs? We include a description of the personal experience of one of us in box 11.1.

### Jackpot! Two Jobs, One School

The best scenario occurs when the same institution hires you both. This can occur in various ways. Perhaps the most common strategy of dual-career couples is for each partner to apply separately for jobs that are suitable. Then, sometime during the process, when one of you obtains a job offer, you negotiate a second tenure-track position for your partner. The big payoff for this strategy is the coveted "two jobs in one place" bonanza. Unlike those pursuing the commuter marriage option (see below), you do not have to wait for a matching pair of jobs to be advertised, and you do not settle for a shared position. Because you can implement this strategy for a job offer to either partner, you maximize the number of jobs that are "in play" with this strategy. If both you and your partner are highly desirable (a likely situation, since someone who is as smart, talented, and ambitious as you are will surely choose an equally smart, talented, and ambitious partner), this approach can actually be a pleasant surprise for the university that is recruiting you. We have seen this strategy work for spouses hired within the same department, as well as for spouses hired in different departments on the same campus.

On the downside, this approach creates the most complicated application, and it can result in tough negotiations. This strategy is always a calculated risk. If you pursue this option, be prepared for some pointed negotiations. The chair may give you an unequivocal "No, we cannot do that." If so, are you prepared to walk away or is one job better than zero? Alternatively, the chair may offer something less than you wanted, perhaps the offer of a temporary position and a pledge to try to create a position. Do not pursue the "negotiate two jobs" strategy in any form until you and your partner agree on what you are and are not prepared to accept.

**Share and Share Alike**

Many scientists meet their spouse or partner while in graduate school, so it is increasingly common for partners to work in the same scientific discipline. This can prompt your friends to ask how you can stand being together twenty-four hours a day, but it also opens another option for applying for jobs. Some dual-career couples opt to apply for positions together and try to share the position. The mechanics of this are straightforward. You both state in your application cover letters that you are applying to share the position advertised. The interview is also simplified because everyone involved is clear on how you intend to handle the dual-career situation.

There are some real benefits to this approach. Most importantly, you and your partner can pursue a career at the same university, thus avoiding the pitfalls of a long-distance relationship. Furthermore, there is no need to decide which career will take a backseat because you are sharing a position as equal partners. Given that two people are sharing one position, this option also creates the possibility that each of you may have a bit more free time to deal with child care or other family issues (but see the important caution below). Although it happens very rarely, we can report that there are real couples out there making this option work successfully for them.

However, beware the serious pitfalls of this approach. First, sharing a tenure-track position is still an unusual approach. Applying jointly carries the risk that you will be excluded from serious consideration for some jobs. A department unfamiliar with this approach is likely to set your applications aside as "too complicated" or simply not worth the extra paperwork that the provost will want filled out. Second, hiring departments will be concerned about what happens if you and your partner decide to split up. This will make some departments gun-shy, and the possibility of a breakup is something you should consider before pursuing this option. Will one of you continue to occupy the position? If so, who? Third, an almost inevitable consequence of sharing a position is that each of you will work more than half-time. Although teaching can be equitably divided, job-sharing couples will routinely find themselves spending more hours on research and service than a single faculty member. Because you are drawing only one salary, this consequence of job sharing can create resentment and dissatisfaction with the position. More generally, this raises the possibility that during tenure and promotion decisions you will actually be evaluated (either explicitly or implicitly) on the grounds that you should be producing something more than a single faculty member. We strongly recommend that you not go into

**Box 11.1. One couple's story**

*One of us (RC) is 50 percent of a dual-career couple. Both he and his wife, Michelle, are vertebrate biologists trained in the same Ph.D. program. This first-person account is meant to describe one of the many options for dealing with this challenge, including the advantages and disadvantages that they faced.*

Michelle and I graduated with our Ph.D.'s in August 1989, and Michelle was pregnant with our only child at the time. In fact, in one memorable week we both defended our dissertations and discovered that we were going to be parents. The pregnancy was according to plan. We had decided that it made the most sense for us to start a family immediately after graduate school. Our reasoning was that we would have more time and flexibility available to us as postdocs than we would as tenure-track faculty. Having a child during graduate school seemed too great a strain on our time and budget.

From the beginning Michelle and I pursued the strategy of obtaining one job and then hoping that a second job became available. Thanks to a generous postdoctoral advisor, this strategy worked well at the outset. I was hired as a postdoctoral researcher at Indiana University, and my supervisor immediately hired Michelle to work on data management and analysis in her lab. However, our success was somewhat fortuitous. My salary was actually paid by the university (I taught one course each year); this gave my advisor the flexibility to pay Michelle a salary from her grant. Over time Michelle was able to work flexible hours through pregnancy and early child care, and she eventually was able to get back into the field to do research. Thus, both Michelle and I were able to come out of our postdoctoral work with good publications. On the downside, the sacrifices involved in this strategy fell primarily on Michelle. Time devoted to pregnancy and child care took more time from her postdoctoral work than from mine, and she had to change her research emphasis (from small mammal ecology to avian behavioral ecology) in order to do postdoctoral work.

After a couple of years as postdocs, we both began applying for appropriate jobs. We applied independently, and we did not mention a dual-career situation in our cover letters. We reasoned that we wanted to maximize the chance that we landed at least one job, and dual-career issues could be pursued later. I should point out that we were targeting midsize regional universities, which in our estimation would be unlikely to have the financial flexibility to create a second job for a spouse (i.e., we did not believe that asking for a second job as part of the application or negotiations would be fruitful). My subsequent experience in midsize state schools supports this assumption.

I was eventually hired as a tenure-track assistant professor at Ball State University in 1992. There were no other permanent job opportunities in the Department of Biology at the time, but Michelle was immediately hired to teach anatomy and physiology as a temporary assistant professor. This had the advantage of providing a firm paycheck (as opposed to the soft money of grant support), and Michelle was able to teach—an activity she enjoys. However, to accommodate our professional situation, Michelle was now making a career choice that would begin to limit her research time and that would have ramifications in the future. Nevertheless, she was still able to publish on her postdoctoral work, as well as get into the field in the summer.

Michelle and I began applying for new jobs because we hoped to move back to the Southeast. Again, she and I applied independently. I accepted a new tenure-track position at Georgia Southern University in 1995, and Michelle was immediately hired to teach introductory courses in the same department. Although her position was "temporary," Michelle was hired full-time and her position proved to be available long term. Given that we did not anticipate moving again, our strategy meant that we were now dependent on tenure-track jobs becoming available at GSU that Michelle would be qualified for. In the meantime, Michelle expanded her teaching activity by acting as co-PI on a number of grants to fund teacher training and outreach. She won a university award for teaching excellence.

Our story has a happy ending. In 2006 Michelle accepted a tenure-track job, and we both now have permanent positions in the same department—the Holy Grail for most dual-career couples. There are two important points concerning her success. First, she was hired in a position that emphasized biology education. Thus, by showing flexibility and creativity as a temporary professor, Michelle eventually carved out an alternative career path (although she will now have the opportunity to get back to her original research interests as well).

Second, although Michelle landed a tenure-track job in 2006, she was not a successful candidate for a position that was open shortly after we arrived at GSU. This disappointment is always a possibility for couples following our strategy, and you must be prepared to deal with the consequences. If you or your partner is not offered an open position, will this create such bitterness that you can't work in the department any longer, or will you be able to move past this event and look for opportunities in the future? Michelle was able to do the latter and succeed in a second job search.

It should be clear from this narrative that the key to success in our career strategy was Michelle's patience and flexibility. For any couple that pursues this strategy, one partner must be willing to accept the uncertainty that goes with moving to new professional situations with no job in place. Nevertheless, Michelle and I have been able to work together in the same department for a span of sixteen years. We count this as a success.

a job-sharing situation without having explicit discussions with your chair concerning tenure expectations. What happens if one of you is an excellent teacher and the other sends students fleeing? If you do not publish jointly, what happens if one of you has ten publications and the other has two?

### The Commuter Marriage

The last, and for most people the least desirable, option is to accept a commuter marriage. Assume that you will both have the good fortune of getting job offers, even if they are on separate coasts, and plan on earning lots of frequent-flier miles over the next few years as you pursue your commuting partnership. The principal benefit of this approach is that you and your partner can both hold full academic positions and be free to pursue your academic careers unencumbered by your partner's situation. Another benefit is that you do not have to worry about dealing with a touchy dual-career situation in your application cover letter or during the interview. For all intents and purposes, your spouse/partner does not exist in the context of your job search. This is clearly the simplest option on paper.

In our opinions, however, searching for jobs independently defeats the very purpose of sharing your life with someone. Choosing this option will place strains on your ability to maintain a relationship (in proportion to the distance between partners). Based on our experience with couples who have followed this strategy, it is most likely to be successful over short distances or for brief periods (e.g., it can work well during your postdoc years).

### When to Announce Your Intentions

Once you and your partner have agreed to a strategy, you need to decide when to let your prospective employer know. There are three approaches: (1) in the cover letter, (2) during the job interview, or (3) after you have received an offer. A short article by Jennifer Thaler and her partner, Anurag Agrawal, lays out the costs and benefits of each of these strategies (Agrawal and Thayer 2003). Here we briefly summarize the issues to consider.

The most open approach is to announce your intentions in your cover letter or during the interview. This provides the department chair an early opportunity to lay the groundwork for coaxing an additional position out of the dean, provost, and so on. Many department chairs, particularly at major research institutions, will tell you that they prefer to have this information right up front. However, to do so also provides search committees maximal opportunity to bump you from contention in order to avoid the complications of finding a second position. The motivations for doing this are complex. Some departments will simply shy away from novel hiring scenarios ("We've never done it that way before!"). Other departments may be willing but will be under financial constraints that legitimately prevent them from offering a second position. As a general rule, we believe that the larger and more research-oriented a university is, the more likely this strategy is to be successful and the more likely you are to benefit from announcing your intentions early.

Of course, it is possible that two tenure-track jobs may become available at the same university at the same time. This is the jackpot for those pursuing this strategy. In this case, we recommend that you mention in your cover letter that you are applying for position A and your partner is applying independently for position B. If either of you is competitive for one position, we believe this is likely to ensure that your partner at least gets serious attention. If two positions are available, we do not believe there is any serious downside to announcing that your partner is in the pool for another job.

You may also choose to let the department know about your situation during the interview process, once they have had the opportunity to discover that you are just as terrific as your letters of reference claimed. It is best to let the department head or chair of the search committee know first, but if it comes up in discussion with others in the department, don't sweat it. Whether you are looking for a second position in the same department or in a different department, it will help to have your partner's application available so the department head has all the necessary information.

If you are trying to maximize the number of job offers and you are anxious not to scare away potential suitors, you may decide to withhold mention of a job-hunting spouse until you have an offer in hand. This gives the department little opportunity to lay the groundwork for a second position (and it could antagonize the chair), but it also provides no opportunity for your spousal situation to diminish your chance for an offer.

Finally, if you are willing to accept less than a tenure-track job for your partner, you have crossed the line into what could be considered a final strategy: have one partner accept a job and then find the other partner the best option after the fact. This option prevents the job search or negotiations from being complicated by dual-career issues, but it provides no guarantee that your spouse will find acceptable work. In addition, we have seen cases where it creates tension between the faculty member who was hired and the department that doesn't seem to be doing enough to find the unemployed partner a job.

No matter which strategy you pursue as a dual-career couple, be prepared for the fact that the department to which you are applying may be way ahead of you. The fact that you and your spouse are both in science is probably widely known in your field (you may publish together). Furthermore, members of search committees often do a little investigative journalism ("Hey, Jan, a postdoc in your department applied for our job. What can you tell me about him/her?"). Thus, the fact that your partner is also a scientist may be known in the department regardless of when you divulge the information. This may lead to some careful questioning during the interview. Our advice to you is be prepared to discuss your spousal situation at any time, regardless of whether or not you planned to broach this subject.

## Women and Negotiation

As academics, most of us are not accustomed to people offering us large sums of money. We are even less experienced at asking for more than we are initially offered. But that can be a central part of the negotiation process (see chapter 10), and one that many women find especially challenging. According to Carol Frohlinger, the co-founder of Negotiating Women (http://www .negotiatingwomen.com) and co-author of *Her Place at the Table: A Woman's Guide to Negotiating Five Key Challenges to Leadership Success*, unique issues often arise for women when they negotiate the terms of a new job.

There are disparities in salaries across all types of universities and across all ranks. According to the American Association of University Professors (AAUP), the average academic woman earns 80 percent of the average man. We worry that this disparity begins at the negotiation table, when women are less likely than men to press hard to be paid what they are worth. What can you do to improve your chances to get your fair share?

- Do your homework. Learn as much as you can to determine what others with credentials similar to yours have received. Think broadly about what is important to you—the compensation package as well as other things that will enable you to be successful (e.g., lab space, a reasonable teaching load, etc.). Personal networks are invaluable as you scout for informal intelligence.
- Get out of your own way. Don't be reluctant to advocate for yourself in a style that is comfortable for you. Nobody else will do it for you.
- Consider your alternatives. What are your other choices if this offer doesn't meet your needs? How attractive are those choices? (As the Gambler advises, "Know when to walk away"). Consider the other party's alternatives. Often their alternatives aren't as good as we might think they are.
- One final thought: Think ahead of time not only about what you will say, but anticipate the responses you may hear. Enlist a friend and practice out loud.

## Domestic Partners

As we discussed earlier, if your partner is an academic, finding the right job can become even more difficult. But if your partner is the same sex as you, even if he or she is not an academic, you face a unique set of challenges. For married couples, a non-academic spouse of a college professor can expect to gain a range of benefits. These might include child care, family leave, retirement plans, insurance policies, tuition remission, and use of campus facilities. The Human Rights Campaign Foundation (http://www.hrc.org) found that nearly three hundred campuses across the country offered some form of benefits for domestic partners. But many do not. As you consider any job offers that you have, you may want to find out if your partner is allowed to use the sports facilities, to obtain health care through your university-based plan, or to attend classes at reduced cost.

### Resources

There are fantastic resources out there to help you sort through issues that women face during the job search. A good website to start with is that of the Association for Women in Science (AWIS, http://www.awis.org). The organization has a "useful links" page pointing to a large number of organizations for women in science, other women's organizations, and sites focused on career development.

The AAUP has focused extensively on issues of interest to women in science. You can find their policy statement and links to many other relevant sites at http://www.aaup.org.

The National Science Foundation has taken a keen interest in increasing the representation of women at all levels of science. The NSF's ADVANCE program has provided funds to universities throughout the country to increase the number of women pursuing academic careers in science and engineering (http://www.nsf.gov).

Finally, for an entertaining look at issues faced by women in academia at all stages, we recommend Emily Toth's book *Ms. Mentor's Impeccable Advice for Women in Academia*.

# AFTERWORD

We hope that after reading this book, you feel better prepared for the up-coming academic job season. If you are just setting out on this journey, we hope that you have a better sense of what you can do years ahead of time to make your life easier once you put yourself out there on the market.

While the three of us went through the process some years ago, we all still remember the challenges, the anxieties, the excitement, and the stimulation of seeking a tenure-track position. We also recognize that we were pretty naive as to what exactly this job was all about.

We want to emphasize that we are not the only individuals who could have written a book that provides some insight into the path to becoming a university professor and the lifestyle once you get there. In fact, one colleague told us that we were the only ones who were dumb enough to do so! Now that we're done, how will we measure success? As empirical biologists, we would love to set up a case-control study comparing the success and happiness of a cohort of male and female academics throughout their careers, some who read this book and others who did not.

Realistically, we recognize that this is one among many sources that you might use to help you navigate these waters. There are many other books on how to find your way into academia and how to succeed once you get there. An impressive list of these books can be found at http://www.phds.org/.books (phds.org is a site devoted entirely to providing career resources for students in science, math, and engineering). There are countless other websites that offer information, as well as blogs and discussion groups about the academic job search. The journals *Science* and *Nature* both sponsor online sites (http://sciencecareers.sciencemag.org and http://www.nature.com/naturejobs, respectively) that not only advertise jobs, but also provide regularly updated information for students and postdocs in science. The *Chronicle of Higher Education* advertises a broader range of academic jobs, including part-time replacement jobs as well as those in more teaching-intensive schools. The website also regularly posts articles by and about students and postdocs (http://chronicle.com).

Finally, we hope that you will come to see this entire process as a wonderful adventure. For those willing to work hard, academia is one of the few jobs to provide intellectual stimulation, diversity of challenges, and the freedom to choose what you do and how and when you do it. So try to have as much fun as you can with this entire process.

# REFERENCES CITED

Agrawal, A., and J. Thayer. 2003, March 7. "Solving the Two-Body Problem." ScienceCareers
   .org.

Barker, K. 2002. *At the Helm: A Laboratory Navigator.* Cold Spring Harbor, NY: Cold Spring
   Harbor Laboratory Press.

Edgerton, R. 1995. *The Teaching Portfolio: Capturing the Scholarship in Teaching.* Washington,
   DC: American Association for Higher Education.

Edgerton, R., P. Hutchings, and K. Quinlan. 1993. *The Teaching Portfolio: Capturing the Scholarship in Teaching.* Washington, DC: American Association for Higher Education.

Golde, C. M., and T. M. Dore. 2001. "At Cross Purposes: What the Experiences of Doctoral
   Students Reveal about Doctoral Education." Philadelphia: Pew Charitable Trusts.
   www.phd-survey.org.

Goldsmith, J. A., J. Komlos, and P. S. Gold. 2001. *The Chicago Guide to Your Academic Career:
   A Portable Mentor for Scholars from Graduate School through Tenure.* Chicago: University
   of Chicago Press.

Kass, L. R. 1999. *The Hungry Soul: Eating and the Perfecting of Our Nature.* Chicago: University
   of Chicago Press.

Kolb, D. M., J. Williams, and C. Frohlinger. 2004. *Her Place at the Table: A Woman's Guide to Negotiating Five Key Challenges to Leadership Success.* San Francisco: Jossey-Bass.

Robbins-Roth, C. 2005. *Alternative Careers in Science: Leaving the Ivory Tower.* New York: Academic Press.

Seldin, P. 2004. *The Teaching Portfolio: A Practical Guide to Improved Performance and Promotion/Tenure Decisions.* Bolton, MA: Anker .

Smaglik, K. 2001. "Demographic Shifts." *Nature* 414:3.

Toth, E. 1997. *Ms. Mentor's Impeccable Advice for Women in Academia.* Philadelphia: University of Pennsylvania Press.